# BASICS OF RELIABILITY AND RISK ANALYSIS

## Worked Out Problems and Solutions

**SERIES ON QUALITY, RELIABILITY AND ENGINEERING STATISTICS**



---

*Published*

Vol. 1: Software Reliability Modelling
*M. Xie*

Vol. 2: Recent Advances in Reliability and Quality Engineering
*H. Pham*

Vol. 3: Contributions to Hardware and Software Reliability
*P. K. Kapur, R. B. Garg and S. Kumar*

Vol. 4: Frontiers in Reliability
*A. P. Basu, S. K. Basu and S. Mukhopadhyay*

Vol. 5: System and Bayesian Reliability
*Y. Hayakawa, T. Irony and M. Xie*

Vol. 6: Multi-State System Reliability
Assessment, Optimization and Applications
*A. Lisnianski and G. Levitin*

Vol. 7: Mathematical and Statistical Methods in Reliability
*B. H. Lindqvist and K. A. Doksum*

Vol. 8: Response Modeling Methodology: Empirical Modeling for Engineering and Science
*H. Shore*

Vol. 9: Reliability Modeling, Analysis and Optimization
*H. Pham*

Vol. 10: Modern Statistical and Mathematical Methods in Reliability
*A. Wilson, S. Keller-McNulty, Y. Armijo and N. Limnios*

Vol. 11: Life-Time Data: Statistical Models and Methods
*J. V. Deshpande and S. G. Purohit*

Vol. 12: Encyclopedia and Handbook of Process Capability Indices:
A Comprehensive Exposition of Quality Control Measures
*W. L. Pearn and S. Kotz*

Vol. 13: An Introduction to the Basics of Reliability and Risk Analysis
*E. Zio*

Vol. 14: Computational Methods for Reliability and Risk Analysis
*E. Zio*

Vol. 15: Basics of Reliability and Risk Analysis: Worked Out Problems and Solutions
*E. Zio, P. Baraldi and F. Cadini*

Series on Quality, Reliability and Engineering Statistics Vol. 15

# BASICS OF RELIABILITY AND RISK ANALYSIS

## Worked Out Problems and Solutions


Enrico Zio

École Centrale Paris et Supelec, France
&
Politecnico di Milano, Italy

Piero Baraldi

Politecnico di Milano, Italy

Francesco Cadini

Politecnico di Milano, Italy



World Scientific

NEW JERSEY • LONDON • SINGAPORE • BEIJING • SHANGHAI • HONG KONG • TAIPEI • CHENNAI

*Published by*

World Scientific Publishing Co. Pte. Ltd.

5 Toh Tuck Link, Singapore 596224

*USA office:* 27 Warren Street, Suite 401-402, Hackensack, NJ 07601

*UK office:* 57 Shelton Street, Covent Garden, London WC2H 9HE

**British Library Cataloguing-in-Publication Data**
A catalogue record for this book is available from the British Library.

**Series on Quality, Reliability and Engineering Statistics — Vol. 15**
**BASICS OF RELIABILITY AND RISK ANALYSIS**
**Worked Out Problems and Solutions**



ISBN-13 978-981-4355-03-2
ISBN-10 981-4355-03-8

Printed in Singapore.

# Contents

## Chapter 1

# Introduction

Reliability and safety are fundamental attributes of any modern technological system. To achieve this, diverse types of protection barriers are placed as safeguards from the hazard posed by the operation of the system, within a *multiple-barrier* design concept. These barriers are intended to protect the system from failures of any of its elements, hardware, software, human and organizational.

Correspondingly, the quantification of the probability of failure of the system and its protective barriers, through reliability and risk analyses, becomes a primary task in both the system design and operation phases.

This exercise book serves as a complementary tool which in support to the methodology concepts introduced in the books "*An introduction to the basics of reliability and risk analysis*" and "*Computational methods for reliability and risk analysis*" by Enrico Zio, in that it gives an opportunity of familiarizing with the applications of classical and advanced techniques of reliability and risk analysis.

The parallelism between theory and practice is strengthened by the structure of the exercise book, which is divided in nine Chapters corresponding to those of the methodological book "*An introduction to the basics of reliability and risk analysis*", and an additional Chapter corresponding to the first Chapter of the other book "*Computational methods for reliability and risk analysis*".

Chapter 2 introduces the definition of risk in qualitative terms and its translation into quantitative terms.

Chapter 3 treats the techniques for the identification of the hazards associated to a given system subject to risks. The output expected from these activities consists of a list of the sources of risks which can give rise to significant dangers. The techniques applied in this Chapter allow qualitative analyses of the system and its functions within a systematic framework of procedures, such as the HAZOP and FMECA Tables.

Chapter 4 formulates reliability and risk problems in terms of basic probability laws, e.g. the theorem of total probability, the definition of conditional probabilities, the probability distributions of random variables, discrete (Binomial, Poisson, Geometric, etc.) and continuous (Exponential, Gaussian, Lognormal, etc.).

Chapter 5 is centered on the concept of reliability as quantitative indicator of the performance of systems which must satisfy a specified mission within an assigned period of time. The computation of the reliability of simple systems of components (series, parallel, stand-by, etc.) is performed by analytical methods.

Chapter 6 deals with the concept of availability, as quantitative indicator of the ability of systems undergoing maintenance of fulfilling the assigned mission at any specific moment of time.

Chapter 7 presents a number of exercises of application of fault tree analysis, which is a systematic, deductive technique for developing the causal relations among events leading to a given undesired event. Familiarity with these techniques is necessary for dealing with complex, multi-component systems for which failure data cannot be collected and a statistical failure analysis is therefore not possible.

Chapter 8 presents exercises on the application of the event tree analysis for identifying the accident sequences which can generate from an initiating event of system failure; the arising sequences are then quantified in terms of their occurrence probability.

The exercises of Chapter 9 illustrate the way that information on the lifetime distribution of a component can be obtained on the basis of the results of "lifetime tests".

Finally, Chapter 10 introduces the basics of the Markov approach to system modeling for reliability and availability analysis. The stochastic process of evolution of the system in time is described through the definition of system states, the possible transitions among these states and their probability of occurrence. The process can be mathematically described in terms of a system of probability equations which can be solved analytically or numerically.

The realization of this book would have not been possible without the support of the many master and Ph.D. students working in our research group. Many thanks are also due to Sara Bastiani and Filippo Belloni, for their careful and precise editing work.

## Chapter 2

# Basic concepts of safety and risk analysis

### 2.1 Risk definition

Give the proper definition of risk and discuss its implications with respect to design, management and regulation of hazardous systems.

### Solution

An informative and operative definition of risk should allow answering the three fundamental questions of any risk analysis:

— Which sequences of undesirable events transform the hazard into an actual damage?
— What is the probability of each of these sequences?
— What are the consequences of each of these sequences?

The answers to these questions lead to a definition of risk in terms of a set of triplets:

$$R = \{\langle s_i, p_i, x_i \rangle\}$$

where $s_i$ is the sequence of undesirable events leading to damage, $p_i$ is the associated probability and $x_i$ the consequences. Thus, the outcome of a risk analysis is a list of scenarios, such as the one in Table 2.1, which represents the risk.

Table 2.1.

| Sequence | Probability | Consequence |
|---|---|---|
| $s_1$ | $p_1$ | $X_1$ |
| $s_2$ | $p_2$ | $X_2$ |
| ... | ... | ... |
| $s_n$ | $p_n$ | $x_n$ |

On the basis of this information, the designer, the manager and the regulator, can act effectively so as to reduce the risk.

## 2.2 HAZOP and FMEA

Briefly illustrate the HAZOP and FMEA methods for hazard analysis, discussing their differences.

### Solution

1. Failure mode and effect analysis (FMEA)

This is a qualitative method, of inductive nature, which aims at identifying those failure modes of the components which could disable system operation or become initiators of accidents with significant external consequences.
The analysis proceeds as follows:

(i) Decompose the system into functionally independent subsystems; for each subsystem identify the various operation modes (start-up, regime, shut-down, maintenance, etc.) and its configurations when operating in such modes (valves open or closed, pumps on or off, etc.).
(ii) For each subsystem in each of its operation modes, compile a table such as Table 2.1, without neglecting any of the subsystem components. The table for a component should

include its failure modes and the effects that such failure has on other components, on the subsystem and on the whole plant.

The analysis takes into account only the effects of single-failures, except for the case of stand-by components, whose failure effects are obviously considered only in case of intervention due to the failure of the main component. Then, in general, there is no indication of the risk associated with multiple or common cause failures.
To ensure a coherent analysis, the analyst must be sure that similar components are given the same failure modes, with same probability values.

2. Hazard and operability analysis (HAZOP)

HAZOP is a qualitative methodology which embraces deductive aspects (search for causes) and inductive aspects (consequences analysis) with the objective of identifying the initiating events of undesired accident sequences. Contrary to FMEA, which is mainly based on the structural/hardware aspects of the system, HAZOP looks at the processes which are undergoing in the plant. Indeed, the method, initially developed for the chemical process industry, proceeds through the compilation of tables (such as Table 2.2) which highlight possible process anomalies and their associated causes and consequences.

The analysis proceeds as follows:

(i) Decompose the system into functionally independent process units (reaction unit, storage unit, pumping unit, etc.); for each process unit identify the various operation modes (start-up, regime, shut-down, maintenance, etc.).
(ii) For each process unit and operation mode, identify the potential deviations from the nominal process behavior. In order to do this, one should:

a. specify all the unit incoming and outgoing fluxes (energy, mass, control signals, etc.) and the characteristic process variables (temperature, flow rate, pressure, concentrations, etc.);
b. write down the various functions that the unit is supposed to fulfill (heating, cooling, pumping, filtering, etc.);
c. apply keywords (low, high, no, reverse, etc.) to the previously identified process variables and unit functions, so as to generate deviations from the nominal process regime.

(iii) For each process deviation (qualitatively) identify its possible causes and consequences. For the consequences, include effects also on other units: this allows HAZOP to account also for domino effects among different units.

Commercial software tools to guide HAZOP are available.

UNIT
OPERATION MODE

Table 2.2. Typical HAZOP table

| **Keyword** | **Deviation** | **Cause** | **Consequence** | **Hazard** | **Actions needed** |
|---|---|---|---|---|---|
| More | More Temp. | Additional Thermal resistance | Higher Pressure in tank | Release Due to Overpressure | Install high temperature warning and pressure relief valve |

## Chapter 3

# Methods for hazard identification

### 3.1 A small external pool

In the small external pool shown in Figure 3.1, the re-circulating water is filtered and purified before being released in the swimming pool.
Perform FMECA and HAZOP.

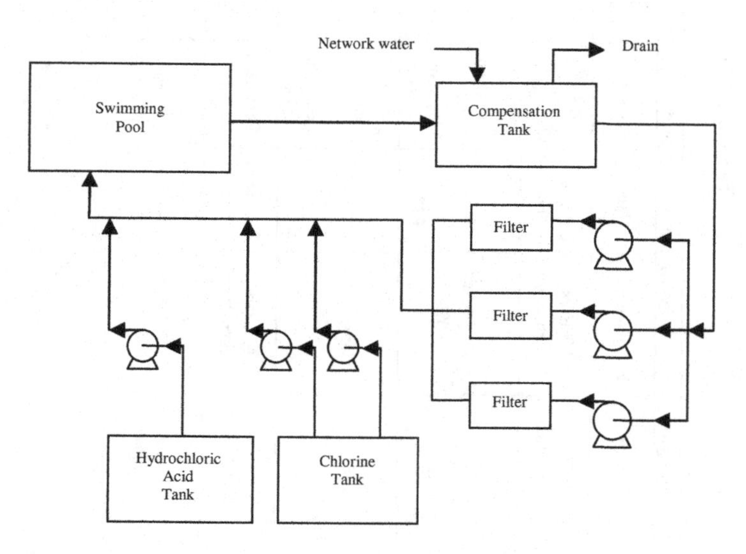


Figure 3.1. Small external pool

## Solution

Table 3.1. FMECA

| Description of component | | Description of the failure | | | Effect of failure | | Criticall y class | Frequency |
|---|---|---|---|---|---|---|---|---|
| **Component** | **Function** | **Failure mode** | **Cause** | **Detection** | **To other components** | **To main function** | | |
| Hydrochloric tank | Ph correction | Measure broken down | Deteriorat ion | Inspection or operation | Swimming pool without correct PH | Bad water quality | Marginal | Reasonable probable |
| Hydrochloric tank pump | Circulate the hydrochloric acid | Pump broken down | - Blockade - No power | -Visual detection - Complaint of users | Swimming pool without correct PH | Bad water quality | Marginal | Remote |
| Filters | Water filtrate | Clogging of filter | Deteriorat ion | -Visual detection - Complaint of users | Swimming pool without filtrate | Bad water quality | Marginal | Probable |
| Chlorine tank | Disinfection | Measure broken down | Deteriorat ion | Complaint of users | Swimming pool not disinfected | Water quality not guarantied | Critical | Reasonable probable |
| Pumps | Pumping water | Pump broken down | - Blockade - No power | Visual detection | No circulation of water | Water quality not guarantied | Marginal | Remote |

Table 3.2. HAZOP analysis

| Guide word | Deviation | Possible causes | Consequences | Proposed measures |
|---|---|---|---|---|
| NO/NOT | There isn't re-circulation of water | Failure of power | Bad water quality | Install generator |
| | | Failure of pumps | Bad water quality | Preventive maintenance of pumps |
| | No chlorine introduced | Lack of chlorine | Bad water quality (water not disinfected) | Install level sensor in chlorine tank |
| | | Failure of pumps | Bad water quality (water not disinfected) | Preventive maintenance of pumps |
| | No hydrochloric acid introduced | Lack of hydrochloric acid | Bad water quality (water with inadequate PH) | Install level sensor in hydrochloric tank |
| | | Failure of pump | Bad water quality (water with inadequate PH) | Preventive maintenance of pumps |
| LESS | Insufficient re-circulation of water | Failure of pumps | Bad water quality | Preventive maintenance of pumps |
| | Insufficient chlorine introduced | Measure broken down | Bad water quality (water not disinfected) | More control/inspection of the measure |
| | Insufficient hydrochloric acid introduced | Failure of pumps | Bad water quality (water with inadequate PH) | Preventive maintenance of pumps |
| | | Measure broken down | Bad water quality (water with inadequate PH) | More control/inspection of the measure |

Table 3.2. (Continued)

| | | | | |
|---|---|---|---|---|
| **MORE** | Excess chlorine introduced | Measure broken down | Rash of the mucous membrane | More control/inspection of the measure |
| | Excess hydrochloric acid introduced | Measure broken down | Bad water quality (water not inadequate PH) | More control/inspection of the measure |

## 3.2 Domestic hot water system

In the domestic hot water system shown in Figure 3.2 the control of the temperature is achieved by the controller opening and closing the main gas valve when the water temperature goes outside the preset limits $T_{min} = 140$ F, $T_{max} = 180$ F.

1. Formulate a list of undesirable safety and reliability events.
2. Perform FMECA.
3. Perform HAZOP.

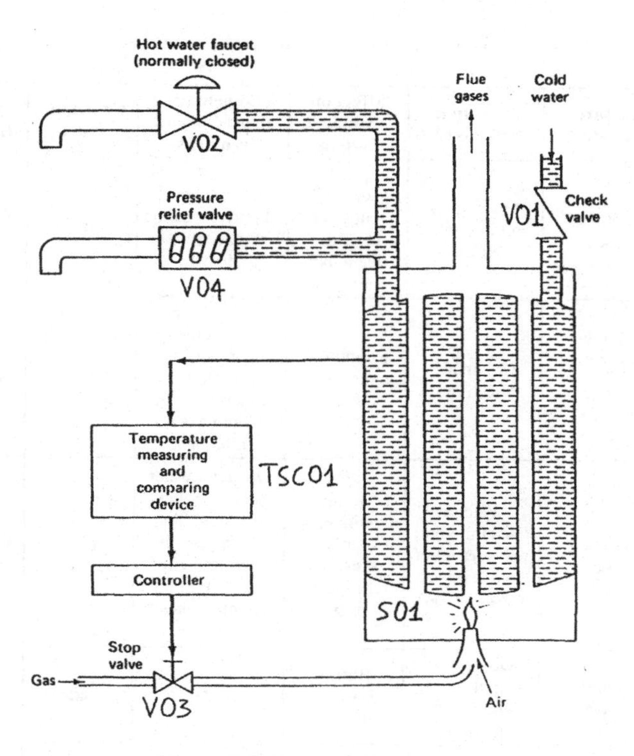


Figure 3.2. Domestic hot water system

**Solution**

1. List of undesirable safety and reliability events

The undesirable safety and reliability events are the following:

- Tank rupture (safety);
- Water too cold (reliability);
- Water too hot (safety/reliability);
- Insufficient water flow (reliability);
- Excessive flow (reliability).

2. FMECA
3. HAZOP

Table 3.3. FMECA analysis

| Component | Failure mode | Detection methods | Effect on whole system | Compensating provision and remarks | Criticality class | Failure frequency |
|---|---|---|---|---|---|---|
| **Pressure relief valve (V04)** | Jammed open | Observe at pressure relief valve | ↑ operation of TS controller; gas flow due to hot water loss | Shut off water supply, reseal or replace relief valve | Safe | Reasonably probable |
| | Jammed close | Manual testing | Rupture of container or pipes | If combined with other component failure, otherwise this failure has no consequence | Critical | Probable |
| **Gas valve (V03)** | Jammed open | Water at faucet too hot; pressure relief valve open (observation) | Burner continues to operate, pressure relief valve opens | Open hot water faucet to relieve pressure. Shut off gas supply. Pressure relief valve compensates. IE1 | Critical | Reasonably probable |
| | Jammed close | Observe at output (water temperature too low) | Burner ceases to operate | / | Safe | Remote |
| **Temperature measuring and comparing device (Tsc01)** | Fail to react to temperature rise above preset level | Observe at output (faucet) | Controller, gas valve, burner continue to function "on". Pressure relief valve opens | Pressure relief valve compensates. Open hot water faucet to relieve pressure. Shut off gas supply. IE2 | Critical | Remote |
| | Fail to react to temperature drop below preset level | Observe at output (faucet) | Controller, gas valve, burner continue to function "off". | / | Safe | Remote |

Table 3.4. HAZOP analysis

| **Guide word** | **Deviation** | **Possible causes** | **Consequence** | **Proposed measures** |
|---|---|---|---|---|
| **MORE** | Gas flow | -Gas valve jammed open; -Control system fails to react to $T>T_{max}$; -Pressure relief valve jammed open. | Possible temperature increase and overpressure in case of a break of the pressure relief valve | Gas manual closing |
| | Water flow | -Pressure relief valve jammed open | Continuous Consumption of water and gas | Water and gas closing |
| | Temperature | -pressure relief valve closed and open breach of gas valve and/or control system | Possible breach due to overpressure | Gas closing Opening of the hot water faucet |
| **LESS** | Gas flow | -Gas valve jammed closed -Control system broken | Breakdown of the system | Substitution of valve and control system |
| | Temperature | -Gas valve jammed closed -Control system fails to react to T<Tmin | Reduction of the system functionality | Substitution of valve and control system |
| **NO** | Control signal | Cables' break down | V03 jammed open or closed. | Substitution of the cables |

## 3.3 Chemical reactor

In the chemical reactor shown in Figure 3.3, reaction of substances A and B leads to the formation of a new substance. In the case where the quantity of B is greater than the quantity of A, there may be an explosion. Perform a HAZOP.

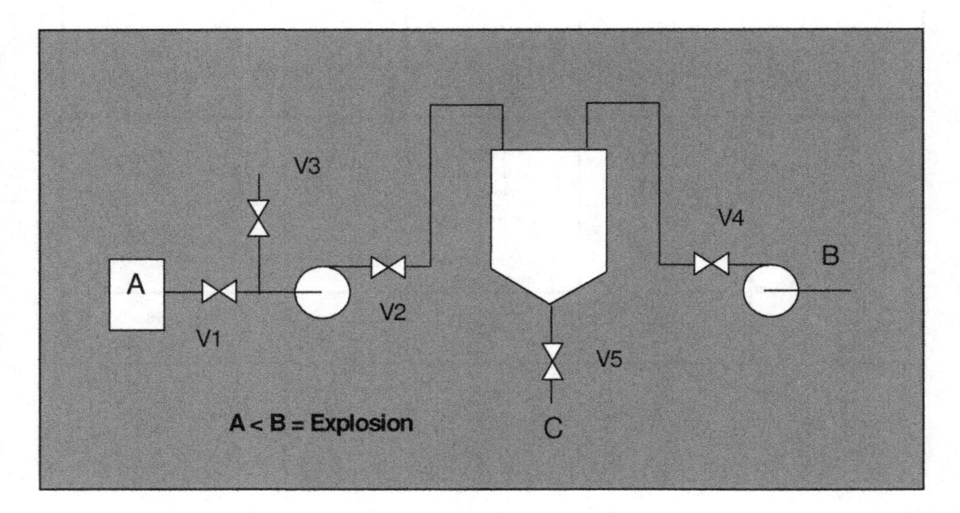


Figure 3.3. Chemical reactor

## Solution

Table 3.5. HAZOP analysis

| Guide Word | Deviation | Possible Causes | Consequences | Proposed Measures |
|---|---|---|---|---|
| **NO/NOT** | No A | Tank containing A is empty.<br>V1 or V2 closed.<br>Pump does not work.<br>Pipe broken | Not enough A = Explosion | Indicator for low level.<br>Monitoring of flow |
| **MORE** | Too much A | Pump too high capacity<br>Opening of V1 or V2 is too large. | C contaminated by A. Tank overfilled. | Indicator for high level.<br>Monitoring of flow |
| **LESS** | Not enough A | V1, V2 or pipe are partially blocked. Pump gives low flow or runs for too short a time | Not enough A = Explosion | See above |
| **AS WELL AS** | Other substance | V3 open – air sucked in | Not enough A = Explosion | Flow monitoring based on weight |
| **REVERSE** | Liquid pumped backwards | Wrong connector to motor | Not enough A = Explosion<br>A is contaminated | Flow monitoring |
| **OTHER THAN** | A boils in pump | Temperature too high | Not enough A = Explosion | Temperature (and flow) monitoring |

### 3.4 Coffee mill

Perform a HAZOP of the coffee mill shown in Figure 3.4.

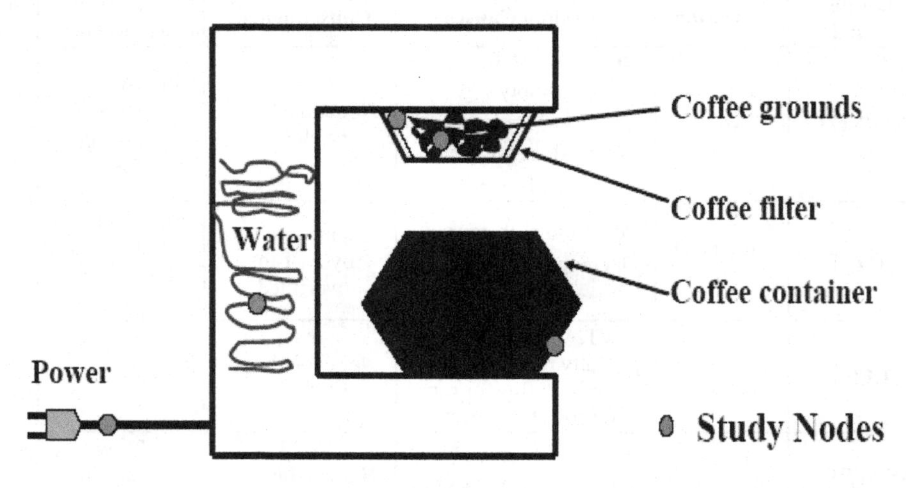


Figure 3.4. Coffee mill

## Solution

Table 3.6. HAZOP analysis

| Guide word | Deviation | Possible causes | Consequences | Proposed measures |
|---|---|---|---|---|
| NO | No flow | No water | No coffee | Check water flow |
| | | Plugged spout | No coffee | Clean spout |
| | | No power | No coffee | Check power |
| | | Basket plugged | No coffee | Clean basket |
| MORE | More flow | Too much water | Pot overflows | Check level |
| LESS | Less flow | Not enough water | Pot not filled | Check level |

## 3.5 Electric fryer

Given the system in the Figure 3.5 perform a recursive HAZOP.

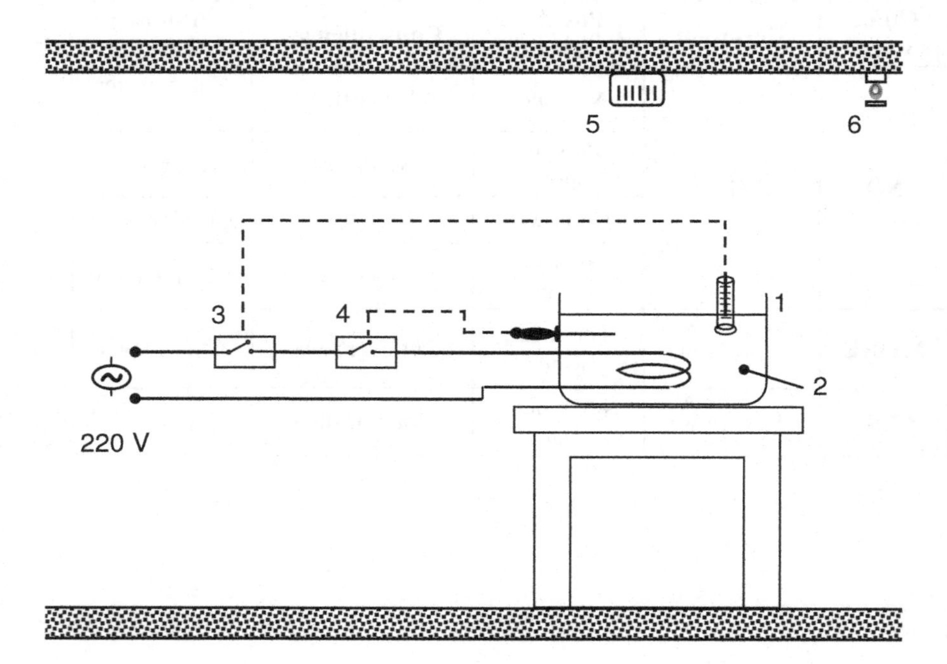


Figure 3.5. Electric fryer

Legend:

1 Electric fryer
2 Oil
3 Thermostat
4 High temperature switch
5 Smoke detector
6 Sprinkler

## Solution

Table 3.6. Recursive HAZOP analysis

| Deviations | Possible causes | Consequences | Proposed measures | | Notes |
|---|---|---|---|---|---|
| | | | Warnings (optic/ acoustic) | Automated protective systems | |
| High temperature | Excessive heat supplied | Very high temperature | | High temperature switch | |
| Very high temperature | High temperature | Boiling oil | Olfactory | | Oil may degrade |
| Boiling oil | Very high temperature | Localized fire | Smoke detector | | |
| Localized fire | Boiling oil | Generalized fire | | Sprinkler | |
| Excessive heat supplied | Thermostat jammed closed | High temperature | | | |

# Chapter 4

# Basics of probability theory for applications to reliability and risk analysis

## 4.1 Compressors failure

Ten compressors, each one with a failure probability of 0.1, are tested independently.

1. What is the expected number of compressors that are found failed?
2. What is the variance of the number of compressors that are found failed?
3. What is the probability that none will fail?
4. What is the probability that two or more will fail?

### Solution

We make use of the binomial distribution $B(k|n,p)$, which in our case gives the probability of observing $k$ failures in $n$ independent tests, when the probability of a failure in one test is $p$:

$$B(k|n,p) = \binom{n}{k} \cdot (p)^k \cdot (1-p)^{n-k}$$

1. Considering the binomial distribution, the expected number of failures $E[k]$ is:

$$E[k] = \mu = \sum_{k=0}^{n} k \cdot B(k|n, p) = n \cdot p = 10 \cdot 0.1 = 1$$

2. The variance $Var[k]$ is:

$$Var[k] = \sum_{k=0}^{n} (k-\mu)^2 \cdot B(k|n, p) = n \cdot p \cdot (1-p) = 1 \cdot (1-0.1) = 0.9$$

3. Probability of no failures:

$$B(k=0|10, p) = \binom{10}{0} \cdot (p)^0 \cdot (1-p)^{10} = 1 \cdot 1 \cdot (1-0.1)^{10} = 0.349$$

4. Probability of two or more failures:

$$\Pr(k \geq 2|10, p) = 1 - B(k=0|10, p) - B(k=1|10, p)$$

$$= 1 - B(k=0|10, p) - \binom{10}{1} \cdot (p)^1 \cdot (1-p)^9$$

$$= 1 - 0.349 - 10 \cdot 0.1 \cdot (1-0.1)^9 = 0.264$$

### 4.2 Spare unit allocation

The allocation of the proper number of spare units for a single-component system is of concern. Assume that if the operating component fails, it is instantaneously replaced by a spare unit, if available. Once the component has failed, it cannot be repaired. The rate of occurrence of components' failures is $1.67\,y^{-1}$, if the component is operating. Assume that the component cannot fail while in the spares depot.

The system designer's goal is to achieve a probability of system operation (reliability) of at least 0.95 at the mission time $T_M$ of one year. How many spare units should be allocated to achieve this goal?

**Solution**

Nomenclature:
$N_{sp}$ : number of spare parts available in the nominal configuration.
$\lambda$ : components' failure rate.
$T_M$ : mission time.

Given the assumptions of immediate replacing and no repair, the system is operating as long as spares are available. In other words, the system fails due to the shortage of spare units. The expression for the probability of system operation (reliability) can be easily obtained by considering that spare shortage occurs upon the $N_{sp}+1th$ failure, when the failed component cannot be further replaced by a spare. In fact, the probability of the system working properly at mission time $T_M$ is the probability of having less than $N_{sp}+1$ failure events between the initial time $T_0 = 0$ and $T_M$.
The probability of having exactly $k$ failures in $(0, T_M)$, when the expected number of failures is $\lambda \cdot T_M$, is given by the Poisson distribution:

$$P(k, T_M | \lambda) = \frac{(\lambda T_M)^k e^{-(\lambda T_M)}}{k!}$$

The system reliability at the mission time $T_M$, $R(T_M)$, is the sum of the probabilities of the mutually exclusive events of having no failures in $(0, T_M)$ one failure in $(0, T_M)$,..., $N_{sp}$ failures in $(0, T_M)$:

$$R(T_M) = \sum_{k=0}^{N_{sp}} \frac{(\lambda T_M)^k e^{-\lambda T_M}}{k!}$$

By trial-and-error, we obtain:

(i) for $N_{sp} = 3$,

$$R(T_M = 1\text{y}) = P[\text{less than } 3+1 \text{ failures in 1 year}] = 0.910$$

(ii) for $N_{sp} = 4$,

$$R(T_M = 1\text{y}) = P[\text{less than } 4+1 \text{ failures in 1 year}] = 0.968$$

Thus, the system reliability requirement $R(T_M = 1y) \geq 0.95$ is satisfied for $N_{sp} = 4$

**4.3 Misprints occurrence in a book**

Consider the occurrence of misprints in a book, and suppose that they occur at the rate of 2 per page.

1. What is the probability that the first misprint will not occur in the first page?
2. Assuming a Poisson process, what is the expected number of pages until the first misprint appears?
3. Comment on the applicability of the Poisson assumptions to the occurrence of misprints or typing errors.

**Solution**

1. For the misprints assume a Poisson distribution with parameter $\upsilon = 2$ misprint / page

$$
\begin{aligned}
&P\{\text{the first misprint will not occur in the } 1^{\text{st}} \text{ page}\} \\
&= P\{0 \text{ misprint in } 1^{\text{st}} \text{ page}\} \\
&= e^{-2} = 0.1353
\end{aligned}
$$

2. $R$ = random variable "number of pages to first misprint"
For $R$ we assume a geometric distribution:

$$P\{\text{first misprint in the } r^{\text{th}} \text{ page}\} = P(R = r) = (1-p)^{r-1}\, p$$

where $(1-p) = P\{\text{no misprints in 1 page}\} = 0.1353$

$$
\begin{aligned}
p &= P\{\text{at least 1 misprint occurring in 1 page}\} \\
&= 1 - 0.1353 \\
&= 0.8647
\end{aligned}
$$

$$E[R] = \frac{1}{p} = \frac{1}{0.865} = 1.16$$

3.

(i) The misprints can actually occur at random and at any page.
(ii) Assuming same conditions in the preparation of every page, the occurrences of misprints over a given section of the book may be acceptably independent of those in any other non overlapping section of the book
(iii) It seems reasonable to assume that the probability of occurrence of a misprint be proportional to the size of the section of the book (in number of pages). The proportionality parameter $\upsilon$ may be assumed constant for all sections, again assuming that all pages have been prepared under the same conditions.
(iv) Still doubts might be raised about the Poisson assumption on the basis that the space over which misprints occur is not

properly a continuum, unless we accept the possibility of having fractional number of pages, in which misprints may occur.

The level of fractioning seems limited by the word entity which cannot further be reduced, so that a real continuum is hard to see.

### 4.4 Peak stresses on a component

Suppose that peak stresses (i.e. stresses which exceed a certain value) occur randomly at an average rate $\lambda$. The probability that a component will survive the application of a peak stress is $(1-p)$ (constant). Show that the reliability of the component is $R(t)=e^{-p\lambda t}$.

**Solution**

$$
\begin{aligned}
&P(r \text{ peak stresses and component survives})\\
&= P(r \text{ peak stresses}) \cdot P(\text{component survives})\\
&\qquad = \frac{(\lambda t)^r}{r!} e^{-\lambda t} (1-p)^r
\end{aligned}
$$

From the theorem of total probability:

$$
\begin{aligned}
R(t) &= P(r=0) \cdot P(\text{component survives} \mid r=0)\\
&\quad + P(r=1) \cdot P(\text{component survives} \mid r=1) + \ldots\\
&= \sum_{r=0}^{\infty} \frac{\lambda t^r}{r!} e^{-\lambda t} (1-p)^r = e^{-\lambda t} \sum_{r=0}^{\infty} \frac{(\lambda t)^r}{r!} (1-p)^r \frac{e^{-(1-p)\lambda t}}{e^{-(1-p)\lambda t}}\\
&= e^{-\lambda t(1-1+p)} \cdot \sum_{r=0}^{\infty} \frac{[(1-p)\lambda t]^r}{r!} \cdot e^{-[(1-p)\lambda t]} = e^{-\lambda p t}
\end{aligned}
$$

## 4.5 Traffic control 1

Suppose that, from a previous traffic count, an average of 60 cars per hour was observed to make left turns at an intersection. What is the probability that exactly 10 cars will be making left turns in a 10 minute interval? Discretize the time interval of interest to approach the problem with the binomial distribution. Show that the solution of the problem tends to the exact solution obtained with the Poisson distribution as the time discretization gets finer.

**Solution**

First we discretize the time interval of one hour into 120 sub-intervals of 30 seconds. Hence, assuming no more than one left turn is possible in any of the 30 seconds interval, the probability $p_{30}$ of a left turn in each 30 seconds interval would be:

$$p_{30} = \frac{60}{120} = 0.5$$

Then, the problem is reduced to find the binomial probability of occurrence of 10 events in 20 trials (20 time intervals of 30 seconds give a time interval of 10 minutes), the probability of an event occurring in each trial being 0.5; thus:

$$P(10 \text{ Left Turns in } 10 \text{ min}) = \binom{20}{10} \cdot (0.5)^{10} \cdot (0.5)^{20-10} = 0.1762$$

This approach holds under the assumption that there is no likelihood of more than one left turns in 30 seconds. In many problems this can be an unrealistic assumption.

Consider now the case that a more accurate solution is required so that time is discretized into 10 seconds intervals. Then the probability $p_{10}$ of observing a left turn in 10 seconds can be estimated as:

$$p_{10} = \frac{60}{360} = \frac{1}{6} = 0.1667$$

Now, the probability of observing exactly 10 left turns becomes:

$$P(10 \text{ Left Turns in 10 min}) = \binom{60}{10} \cdot \left(\frac{1}{6}\right)^{10} \cdot \left(\frac{5}{6}\right)^{60-10} = 0.1370 .$$

Further improvements can be achieved by taking even shorter time intervals. In general, if the time $t$ is divided into $n$ equal intervals, then

$$P(k \text{ events in time } t) = \binom{n}{k} \cdot \left(\lambda \cdot \frac{t}{n}\right)^{k} \cdot \left(1 - \lambda \cdot \frac{t}{n}\right)^{n-k}$$

where $\lambda$ is the average number of events in time $t$. If an event can occur continuously at any time (as in this left turn traffic problem), for $n \to \infty$:

$$P(k \text{ events in time } t) = \lim_{n \to \infty} \binom{n}{k} \cdot \left(\frac{\lambda t}{n}\right)^{k} \cdot \left(1 - \frac{\lambda t}{n}\right)^{n-k} = \frac{(\lambda t)^k}{k!} e^{-\lambda t}$$

which is the Poisson distribution.
In the case of left turns traffic, $\lambda = 1\,\text{min}^{-1}$ so that

$$P(10 \text{ Left Turns in 10 minutes}) = \frac{10^{10}}{10!} \cdot e^{-10} = 0.1251$$

By going back to the values of the probabilities of interest obtained by means of the binomial approach (0.1762 when the binomial parameter is $p_{30} = 0.5$ and 0.1370 when it is $p_{10} = 0.1667$), it is evident that as the time discretization interval decreases, the probability of 10 Left Turns in 10 minutes gets closer to the exact value of 0.1251.

## 4.6 Traffic control 2

Suppose that it is observed that, on average, 100 cars per hour reach an intersection. Also, it has been estimated that the probability for a car to make a left turn is 0.6. What is the probability that exactly 10 cars will be making left turns in a 10 minute interval?

### Solution

Nomenclature:

$\lambda$: rate of arrival of cars at the intersection.
$p$: probability of a left turn at the intersection

On an intuitive basis, an observer counting the number of left turns would experience the same situation as in the exercise "Traffic control 1". Indeed, this observer would register an average of 60 cars per hours $(100 \cdot 0.6)$. Thus, from the point of view of the left-turn observer the process is the same as the one considered in the exercise "Traffic control 1" and we expect the same solution. Note that in this case we have implicitly introduced an effective rate $\lambda' = 60h^{-1} = 1\,\text{min}^{-1}$ for left turns.
Let us now rigorously verify our intuitive result. Note that the observation that $k$ cars turn left at the intersection necessarily descend from the fact that $r(r \geq k)$ cars have reached the intersection and that k cars out of the r have made a left turn. The

number of cars $r \geq k$ can be any and by summing up the probabilities of all these mutually exclusive events, we obtain the probability of k left turns in the time interval $(0,t)$:

$$P\left(k \text{ Left Turns in } (0,t) \middle| \lambda, p\right)$$

$$= \sum_{r=k}^{\infty} p\left(r \text{ cars reaching the intersection} \middle| \lambda\right)$$

$$\cdot p\left(k \text{ cars turn left out of the } r \text{ reaching the intersection} \middle| p\right)$$

$$= \sum_{r=k}^{\infty} e^{-\lambda t} \cdot \frac{(\lambda t)^r}{r!} \cdot \binom{r}{k} \cdot p^k \cdot (1-p)^{r-k}$$

$$= p^k \cdot \frac{e^{-\lambda t}}{k!} \cdot \sum_{r=k}^{\infty} \frac{(\lambda t)^r}{r!} \cdot \frac{r!}{(r-k)!} \cdot (1-p)^{r-k} = [n = r-k]$$

$$= p^k \cdot \frac{e^{-\lambda t}}{k!} \cdot \sum_{n=0}^{\infty} \frac{(\lambda t)^{n+k}}{n!} \cdot (1-p)^n$$

$$= (p\lambda t)^k \cdot \frac{e^{-\lambda t}}{k!} \cdot \sum_{n=0}^{\infty} \frac{(\lambda t(1-p))^n}{n!} = (p\lambda t)^k \cdot \frac{e^{-\lambda t}}{k!} \cdot e^{\lambda t(1-p)}.$$

Thus,

$$P\left(k \text{ Left Turns in } (0,t) \middle| \lambda, p\right) = \frac{(p\lambda t)^k}{k!} \cdot e^{-p\lambda t}$$

which is the Poisson distribution with an effective rate of occurrence $\lambda' = p\lambda$

In our case, $\lambda = \frac{100}{60} \min^{-1} = 1.6667 \min^{-1}$ and $p=0.6$ so that $\lambda' = 1 \min^{-1}$.

$$P(10 \text{ Left Turns in } 10 \text{ minutes}) = \frac{10^{10}}{10!} \cdot e^{-10} = 0.1251$$

which is the same solution found in Exercise 4.25.

**4.7 Aircraft flight panel**

An aircraft flight panel is fitted with two types of artificial horizon indicators. The times to failure of each indicator from the start of a flight follow an exponential distribution with a mean value of 15 hours for one and 30 hours for the other. A flight lasts for a period of 3 hours.

1. What is the probability that the pilot will be without an artificial horizon indication by the end of a flight?
2. The mean time to this event, if the flight is of a long duration?

**Solution**

1. $\upsilon = \frac{1}{\mu}$

For exponential distribution, *P(failure before t)=* $1 - e^{-\upsilon t}$

$$P(\text{system 1 fails before 3 hrs}) = 1 - e^{-3/15}$$

$$P(\text{system 2 fails before 3 hrs}) = 1 - e^{-3/30}$$

*P(no horizon indicator after 3 hrs)*

*=P(system 1 fails and system 2 fails)*

$$= (1 - e^{-3/15}) \cdot (1 - e^{-3/30})$$

$$= 0.01725$$

2. *P(failure before t)* = $\left(1-e^{-t/15}\right)\cdot\left(1-e^{-t/30}=F(t)\right)$

$$MTTF=\int_0^\infty tf(t)dt=\int_0^\infty R(t)dt=\int_0^\infty (1-F(t))dt$$

$$=\int_0^\infty\left(1-1+e^{-t/15}+e^{-t/30}-e^{-t\left(\frac{1}{15}+\frac{1}{30}\right)}\right)dt$$

$$=(-15e^{-t/15}-30e^{-t/30}+10e^{-t/10})\Big|_0^\infty$$

$$=15+30-10=35 \text{ hours}$$

**4.8 Simple system**

Consider a system of two independent components with exponentially distributed failure times. The failure rates are $\lambda_1$ and $\lambda_2$, respectively.
Determine the probability that component 1 fails before component 2.

**Solution**

Component 2 lives on component 1 if, for example, the failure of component 1 occurs at a time $T_1$ within a time interval $(t,t+dt)$ and the failure of component 2 occurs at a time $T_2$ after *t*. The probability of this event is:

$$P\left(T_2>t\middle|T_1=t\right)\cdot f_{T_1}(t)\,dt$$

where

$$f_{T_1}(t)\ dt = \lambda_1 e^{-\lambda_1 t} dt$$

Furthermore, given the assumption of independent components, the conditional probability $P(T_2 \geq t | T_1 = t)$ is equal to $P(T_2 > t) = e^{-\lambda_2 t}$.

Then, all the contributions for any time interval $(t, t+dt)$ have to be summed to give the required probability $P(T_2 > T_1)$:

$$P(T_2 > T_1) = \int_0^\infty P(T_2 > t | T_1 = t) \cdot f_{T_1}(t)\ dt$$

$$= \int_0^\infty e^{-\lambda_2 t} \cdot \lambda_1 \cdot e^{-\lambda_1 t} dt = \lambda_1 \cdot \int_0^\infty e^{-(\lambda_1+\lambda_2)t} dt = \frac{\lambda_1}{\lambda_1 + \lambda_2}$$

NOTE
This result can be easily generalized to a system of n independent components with failure rates $\lambda_1, \lambda_2, ..., \lambda_n$. The probability that component $j$ is the first one to fail is:

$$P(\text{component } j \text{ fails first}) = \frac{\lambda_j}{\sum_{i=1}^{n} \lambda_i}.$$

**4.9 Machine survive period**

A machine has been observed to survive a period of 100 hours without failure with probability 0.5. Assume that the machine has a constant failure rate $\lambda$.

1. Determine the failure rate $\lambda$.
2. Find the probability that the machine will survive 500 hours without failure.
3. Determine the probability that the machine fails within 1000 hour, assuming that the machine has been observed to be functioning at 500 hours.

**Solution**

1. The component is assumed to have constant failure rate, i.e. to have exponentially distributed failure times. The failure rate $\lambda$ can be determined by:

$$P(\,T_2 > 100) = e^{-\lambda \cdot 100} = \frac{1}{2}.$$

Hence,

$$-\lambda \cdot 100 = -\ln 2 \;\Rightarrow\; \lambda = \frac{\ln 2}{100} = 6.9 \cdot 10^{-3}$$

2. The probability that the machine will survive 500 hours without failure, i.e. the reliability at time 500 hours, $R(500)$, is:

$$P(\,T > 500) = R(500) = e^{-\lambda \cdot T} = e^{-\frac{\ln 2}{100} \cdot 500} = 2^{-5} = \frac{1}{32}$$

3. The problem is solved by the rule of conditional probabilities:

$$P(A|B) = \frac{P(A \cap B)}{P(B)}:$$

$$P(T<1000 \mid T>500)=\frac{P(T<1000 \cap T>500)}{P(T>500)}$$

$$=\frac{P(\ 500<T<1000)}{P(T>500)}=\frac{R(1000)\text{ - }R(500)}{R(500)}$$

$$=\frac{2^{-5}-2^{-10}}{2^{-5}}=0.97$$

## 4.10 Television picture tubes

The television picture tubes of manufacturer $A$ have a mean lifetime of 6.5 years and a standard deviation of 0.9 years, while those of manufacturer $B$ have a mean lifetime of 6.0 years and a standard deviation of 0.8 years. What is the probability that a random sample of 36 tubes from manufacturer $A$ will have a mean lifetime that is at least 1 year more than the mean lifetime of a sample of 49 tubes from manufacturer $B$ ?

(Note: The statistical mean of a random sample, typically called "sample mean", is a function of the random sample and it is, therefore, a random variable itself. In fact, if I take another sample, it will in general be different from the previous one and the sample mean will take a different value. It can be shown that the expected value, or theoretical mean, of the sample mean coincides with the expected value of the underlying population distribution from which the sample was drawn whereas the variance of the sample mean turns out to be equal to the variance of the underlying population distribution, divided by the number n of values constituting the sample.)

**Solution**

$X_1$ is normal with mean 6.5 and standard deviation $=\frac{0.9}{\sqrt{36}}=0.15$

$X_2$ is normal with mean 6.0 and standard deviation $= \dfrac{0.8}{\sqrt{49}} = 0.11$

Then $Y = X_1 - X_2$ is also normal with mean $= 6.5 - 6.0 = 0.5$ and standard deviation $= \sqrt{0.15^2 + 0.11^2} = 0.19$

Let $Z = \dfrac{Y - 0.5}{0.19}$ be the standard normal variate. Then,

$$P[Y > 1] = P\left[Z > \frac{1.0 - 0.5}{0.19}\right] = 1 - P\,[\,Z \le 2.63] = 1 - \Phi(2.63) = 0.0043$$

**4.11 Building safety**

In considering the safety of a building, the total force acting on the columns of the building must be examined. This would include the effects of the dead load $D$ (due to the weight of the structure), the live load $L$ (due to human occupancy, movable furniture, and the like), and the wind load $W$.
Assume that the load effects on the individual columns are statistically independent Gaussian variates with

$$\mu_D = 4.2 \text{ kips} \qquad \sigma_D = 0.3 \text{ kips}$$

$$\mu_L = 6.5 \text{ kips} \qquad \sigma_L = 0.8 \text{ kips}$$

$$\mu_W = 3.4 \text{ kips} \qquad \sigma_W = 0.7 \text{ kips}$$

1. Determine the mean and standard deviation of the total load acting on a column.
2. If the strength $R$ of a column is also Gaussian with a mean equal to 1.5 times the total mean force, what is the probability of failure of the column? Assume that the coefficient of

variation of the strength $\delta_R$ is $15\%$ and that the strength and load effects are statistically independent.

**Solution**

1. The combined load $S$ is

$$S = D + L + W$$

which is also Gaussian with

$$\mu_S = \mu_D + \mu_L + \mu_W = 4.2 + 6.5 + 3.4 = 14.1 \text{ kip}$$

And

$$\sigma_S = \sqrt{\sigma_D{}^2 + \sigma_L{}^2 + \sigma_W{}^2} = \sqrt{0.3^2 + 0.8^2 + 0.7^2} = 1.1 \text{ kip}$$

2. Failure of the column will occur when the strength $R$ is less than the applied load $S$. Let $X$ denote the difference $R - S$, namely,

$$X = R - S$$

then $(X < 0)$ represents the failure. Since $R$ and $S$ are independent Gaussian variables, $X$ is also Gaussian with

$$\mu_X = \mu_R - \mu_S = 1.5 \cdot 14.1 - 14.1 = 7.05 \text{ kip}$$

$$\sigma_S = \sqrt{\sigma_R{}^2 + \sigma_S{}^2} = \sqrt{(\delta_R \mu_R)^2 + \sigma_S{}^2}$$
$$= \sqrt{(0.15 \cdot 1.5 \cdot 14.1)^2 + (1.1)^2} = 3.36 \text{ kip}$$

Hence the probability of failure is

$$P(X<0)=\Phi\left(\frac{0-\mu_X}{\sigma_X}\right)=\Phi\left(\frac{-7.05}{3.36}\right)=\Phi(-2.1)=1-0.982=0.018$$

### 4.12 Capacitor

A capacitor is placed across a power source. Assume that surge voltages occur on the line at a rate of one per month and they are normally distributed with a mean value of 100 volts and a standard deviation of 15 volts. The breakdown voltage of the capacitor is 135 volts.

1. Find the mean time to failure ( $MTTF$ ) for this capacitor
2. Find its reliability for a time period of one month

**Solution**

1. Surge occurrence is a poissonian process with parameter $\lambda = 1 month^{-1}$. The random variable $K$ which indicates how many surges will occur in $(0,t)$ is then distributed as:

$$P[k,(0,t)\,|\,\lambda]=e^{-\lambda t}\frac{(\lambda t)^k}{k!}$$

Surge magnitude $M$ is a normally distributed variable with parameters $\mu = 100$ V and $\sigma = 15$ V. Then, denoting $F_M(\cdot)$ the cumulative distribution function of $M$ ,

$$p = P[\text{lethal surge}] = P[M \geq 135 \text{ V}] = 1 - F_M(M = 135 \text{ V}).$$

Recall now that the variable $Z=\dfrac{M-\mu}{\sigma}$ is a standard normal random variable, i.e. its distribution is normal $F_Z(z)\equiv N(0,1)$. For the value $m = 135$ V, the corresponding standard normal realization is $z = \frac{135-100}{15} = 2.33$ and from the standard normal variate probability Table:

$$p = P[\text{lethal surge}] = 1 - F_Z(135) = 1 - 0.9901 = 0.0099$$

Now, the occurrence of $k$ lethal surges in $(0,t)$ derives from the fact that $r$ surges $(r > k)$ have occurred on the line in $(0,t)$ and that $k$ out of these have been lethal. Obviously, the number of surges $r \geq k$ can be any and by summing up the probabilities of all these mutually exclusive events, we obtain the probability of $k$ lethal surges in the time interval $(0,t)$:

$$P[k \text{ lethal surges in } (0,t) \mid \lambda, p]$$

$$= \sum_{r=k}^{\infty} P[r \text{ surges} \mid \lambda] \cdot P[k \text{ out of } r \text{ surges are lethal} \mid p]$$

$$= \sum_{r=k}^{\infty} \frac{(\lambda t)^r e^{-\lambda t}}{r!} \cdot \binom{r}{k} p^k (1-p)^{r-k} = e^{-\lambda t} p^k \sum_{r=k}^{\infty} \frac{(\lambda t)^r}{r!} \cdot \frac{r!(1-p)^{r-k}}{(r-k)!k!}$$

Substituting $n = r - k$ we have:

$$P[\text{lethal surges in } (0,t) \mid \lambda, p] = \frac{e^{-\lambda t}}{k} (p\lambda t)^k e^{-\lambda t(1-p)} = \frac{(\lambda pt)^k}{k!} e^{-\lambda pt}$$

Since the random variable number of lethal surges', i.e. the number of capacitor failures follows a poissonian distribution with rate $\lambda' = \lambda p$, the corresponding random variable failure time $T$ is distributed as an exponential distribution with same failure rate. Hence the mean time to failure is:

$$MTTF = E[T] = \frac{1}{\lambda'} = \frac{1}{\lambda p}$$

2. By definition,

$$R(t) = P[\text{no failure in } (0,t) \mid \lambda, p] = P[k = 0 \text{ in } (0,t) \mid \lambda, p] = e^{-\lambda pt}$$
$$R(1 \text{ month}) = 0.99$$

**4.13 Hazard function**

The hazard function for a device which has a performance characteristic $x$ is $h(x) = \frac{1}{2\sqrt{x}}$.

Derive the expression for

1. The probability density function
2. The cumulative distribution function
3. Calculate the mean of the PDF
4. Which part of the bathtub curve is this hazard function approximating?

**Solution**

1. $h(x) = \frac{f(x)}{1 - F(x)} = \frac{1}{2\sqrt{x}}$ where $f(x) = \frac{dF(x)}{dx}$

Rewriting the above equation, we get

$$\frac{dF(x)}{dx} + \frac{1}{2\sqrt{x}} F(x) = \frac{1}{2\sqrt{x}}$$

Integrating factor,

$$\mu = e^{\int_0^x \frac{1}{2\sqrt{x}}dx} = e^{\sqrt{x}}$$

Therefore,

$$F(x) = e^{-\sqrt{x}} \int_0^x e^{\sqrt{x}} (\frac{1}{2\sqrt{x}}) dx + ce^{-\sqrt{x}}$$

To solve

$$\int_0^x e^{\sqrt{x}} \frac{1}{2\sqrt{x}} dx,$$

let

$$u = \sqrt{x} \Rightarrow du = \frac{1}{2\sqrt{x}} dx$$

Substituting, we get

$$\int e^u du = e^u + c = e^{\sqrt{x}} + c$$

Then

$$\int_0^x e^{\sqrt{x}} (\frac{1}{2\sqrt{x}}) dx = e^{\sqrt{x}} \Big|_0^x = e^{\sqrt{x}} - 1$$

Replacing in the equation of $F(x)$,

$$F(x) = 1 - e^{-\sqrt{x}} + ce^{-\sqrt{x}} = 1 + ce^{-\sqrt{x}}$$

We know that $\lim_{x \to -\infty} F(x) = 0$, and $\lim_{x \to \infty} F(x) = 1$

Since $F(x)$ is not defined for $x<0$, $F(0)=0=1+c \Rightarrow c=-1$

Then $F(x)=1-e^{-\sqrt{x}}$

$$f(x)=\frac{dF(x)}{dx}=-e^{-\sqrt{x}}(-0.5x^{-0.5})=\frac{e^{-\sqrt{x}}}{2\sqrt{x}}$$

2. From part 1, $F(x)=1-e^{-\sqrt{x}}$

3.

$$E[x]=\int_0^\infty xf(x)dx=\int_0^\infty x\frac{e^{-\sqrt{x}}}{2\sqrt{x}}dx \qquad \text{(Note: } x>0\text{)}$$

let

$$u=\sqrt{x}, du=\frac{1}{2\sqrt{x}}dx \Rightarrow E(x)=\int u^2e^{-u}du$$

From Integral Tables,

$$\int x^m e^{ax}dx=\frac{x^m e^{ax}}{a}-\frac{m}{a}\int x^{m-1}e^{ax}dx$$

$$E[x]=-u^2e^{-u}+2\int ue^{-u}du$$

From Integral Tables,

$$\int ue^{-u}du=e^{-u}(-u-1)$$

$$E[x]=-u^2e^{-u}+2e^{-u}(-u-1)=(-u^2-2u-2)e^{-u}$$

Substituting

$$u=\sqrt{x},\Rightarrow E[x]=-(x+2\sqrt{x}+2)e^{-\sqrt{x}}\Big|_0^\infty=0-(-2)=2$$

Therefore, the mean $E[x] = -2$

4. The part of the bathtub curve which the hazard function $h(x) = \dfrac{1}{2\sqrt{x}}$ is approximating is the burn-in or infant mortality part.

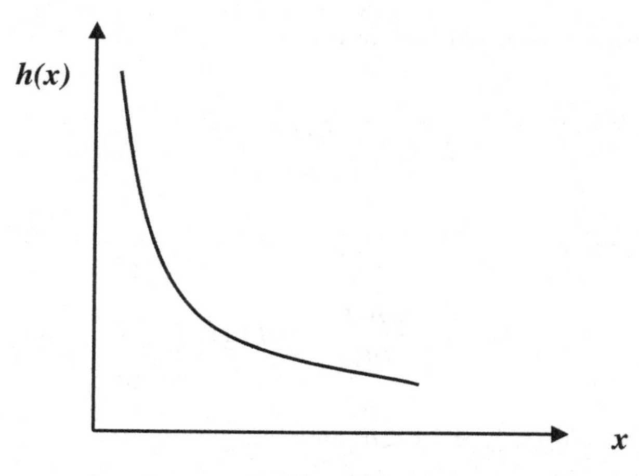


Figure 4.1.

**4.14 Reliability of a device**

1. Calculate the reliability of a device for which the hazard function is defined by:

$$h(t) = \begin{cases} 0 & 0 \le t < a \\ \lambda & a \le t < b \\ \lambda e^{\frac{(t-b)}{c}} & t \ge b, c > 0 \end{cases}$$

2. With $a = 10$ h, $b = 1000$ h, $c = 500$ h, $\lambda = 10^{-4}$ h, calculate the minimum time before the cumulative failure probability $F(t)$ equals 0.2.

**Solution**

1. The definition of hazard function is:

$$h(t) = \frac{f(t)}{1 - F(t)}$$

Then

$$h(t)dt = -\frac{dR(t)}{R(t)} = -d \ln R(t)$$

it is found that $R(t) = 1,\ 0 \le t < a$ and

$$R(t) = R(a)e^{-\int_a^t h(t')dt'} \qquad a \le t < b$$

$$= e^{[-\lambda(t-a)]}$$

Likewise it follows that

$$R(t) = R(b) \cdot e^{-\int_b^t h(t')dt'}$$

$$= e^{-\lambda\left[b-a-c+ce^{\left(\frac{t-b}{c}\right)}\right]} \qquad t \ge b$$

Thus,

$$R(t)=\begin{cases}1 & 0\le t<a\\ e^{-\lambda(t-a)} & a\le t<b\\ e^{-\lambda\left[b-a-c+ce^{\left(\frac{t-b}{c}\right)}\right]} & t\ge b\end{cases}$$

2. For $F(t_{\min})=0.2$, it is obvious that $t_{\min}>a=10$ h, so we first assume $a<t_{\min}<b$, with $b=1000$ h, and use

$$\int_0^{t_{\min}} h(t)dt=\int_0^a h(t)dt+\int_a^{t_{\min}} h(t)dt=-\ln R(t)$$

to find

$$\lambda(t_{\min}-a)=-\ln[R(t_{\min})]=-\ln 0.2$$

After substituting the numerical values, we obtain $t_{\min}=2241$ h. Since $t_{\min}>1000$ h the assumption $a<t_{\min}<b$ was incorrect, so we use

$$\int_0^T h(t)dt=\int_0^a h(t)dt+\int_a^b h(t)dt+\int_b^T h(t)dt=-\ln R(t)$$

to find

$$\lambda\left[(b-a-c)+ce^{\frac{(t_{\min}-b)}{c}}\right]=-\ln 0.2$$

From this equation we conclude that $t_{\min}=1624$ h.

## 4.15 Earthquake-resistant design

The following relationship arises in the study of earthquake-resistant design:

$$Y = ce^X$$

where $Y$ is ground-motion intensity at the building site, $X$ is the magnitude of an earthquake and $c$ is related to the distance between the site and center of the earthquake.
If $X$ is exponentially distributed,

$$f_X(x) = \lambda e^{-\lambda x} \qquad x \geq 0$$

find the cumulative distribution function of $Y$, $F_Y(y)$.

### Solution

We know, $F_X(x) = 1 - e^{-\lambda x}$
Then $F_Y(y) = P\{Y \leq y\} = P\{ce^X \leq y\}$
Since $c \geq 0$:

$$= P\{e^X \leq \frac{y}{c}\} = P\{X \leq \ln\frac{y}{c}\} = F_X\left(\ln\frac{y}{c}\right)$$

$$= 1 - \exp[-\lambda \ln\frac{y}{c}] = 1 - \left(\frac{c}{y}\right)^{\lambda}$$

<u>*Alternative Method:*</u>

$$f_Y(y)dy = f_X(x)dx \qquad x = \ln\left(\frac{y}{c}\right) \qquad \left(\frac{dx}{dy}\right) = \frac{1}{y}$$

$$f_Y(y) = f_X(x)\frac{dx}{dy} = f_X(x)\left(\frac{1}{y}\right) = \lambda e^{-\lambda \ln\frac{y}{c}} \cdot \left(\frac{1}{y}\right) = \frac{\lambda}{y}\left(\frac{y}{c}\right)^{-\lambda} = \lambda c^{\lambda} y^{-\lambda-1}$$

$$F_Y(y) = \int_c^y f_Y(y)dy = \int_c^y \lambda c^{\lambda} y^{-\lambda-1} dy = -\lambda c^{\lambda} \frac{y^{-\lambda}}{\lambda}\Big|_c^y = -c^{\lambda} y^{-\lambda} + 1 = 1 - \left(\frac{c}{y}\right)^{\lambda}$$

## Chapter 5

# Reliability of simple systems

### 5.1 Satellite with two transmitters

Consider a satellite with two transmitters, one of which is in cold standby. Loss of transmission can occur if either both transmitters have failed or solar disturbances permanently interfere with transmission. If the rate of failure of the on-line transmitter is $\lambda$ and the rate of solar disturbances is $\lambda_{cm}$, find:

1. The reliability of transmission.
2. The mean time to transmission failure.

**Solution**

1. Reliability of transmission

   $T =$ transmission failure time

$$P\{\text{trasmission successful at } t\}$$
$$= [P\{\text{Trasmitter 1 does not fail before t}\}$$
$$+ P\{\text{Transmitter 1 fails in } \tau > t \text{ and transmitter 2}$$
$$\text{does not fail from } \tau \text{ to } t\}]$$
$$\times P\{\text{no solar disturbances in } 0,t\}$$

$$R(t) = P(T > t) = \left( e^{-\lambda t} + \int_0^t \lambda e^{-\lambda \tau} e^{-\lambda(t-\tau)} d\tau \right) e^{-\lambda_{cm} t}$$

$$= \left( e^{-\lambda t} + \lambda e^{-\lambda t} \int_0^t d\tau \right) e^{-\lambda_{cm} t} = \left( e^{-\lambda t} + \lambda t e^{-\lambda t} \right) e^{-\lambda_{cm} t}$$

$$= e^{-(\lambda + \lambda_{cm})t} \left( 1 + \lambda t \right)$$

2. The mean time to transmission failure

$$MTTF = T = \int_0^\infty R(t) dt = \int_0^\infty e^{-(\lambda - \lambda_{cm})t} (1 + \lambda t) dt$$

$$= -\frac{1}{\lambda + \lambda_{cm}} e^{-(\lambda + \lambda_{cm})t} \Big|_0^\infty$$

$$+ \left( -\frac{1}{\lambda + \lambda_{cm}} e^{-(\lambda + \lambda_{cm})t} \lambda t \Big|_0^\infty + \frac{1}{\lambda + \lambda_{cm}} \int_0^\infty \lambda e^{-(\lambda - \lambda_{cm})t} dt \right)$$

$$= \frac{1}{\lambda + \lambda_{cm}} + \frac{1}{\lambda + \lambda_{cm}} \left( -\frac{1}{\lambda + \lambda_{cm}} e^{-(\lambda - \lambda_{cm})t} \right)_0^\infty$$

$$= \frac{1}{\lambda + \lambda_{cm}} + \frac{1}{(\lambda + \lambda_{cm})^2}$$

### 5.2 Simple system

Consider a system of two independent components with exponentially distributed failure times. The failure rates are $\lambda_1$ and $\lambda_2$, respectively. Determine the probability that component 1 fails before component 2.

**Solution**

Component 2 lives on component 1 if, for example, the failure of component 1 occurs at a time $T_1$ within a time interval $(t, t+dt)$ and the failure of component 2 occurs at a time $T_2$ after t. The probability of this event is:

$$P(T_2 > t \mid T_1 = t) \cdot f_{T_1}(t)dt .$$

where $f_{T_1}(t)dt = \lambda_1 e^{-\lambda_1 t} dt$. Furthermore, given the assumption of independent components, the conditional probability $P(T_2 > t \mid T_1 = t)$ in equation above is equal to $P(T_2 > t) = e^{-\lambda_2 t}$.

Then, all the contributions as the one considered in the first equation for any time interval $(t, t+dt)$ have to be summed to give the required probability $P(T_2 > T_1)$:

$$P(T_2 > T_1) = \int_0^\infty P(T_2 > t \mid T_1 = t) \cdot f_{T_1}(\mathrm{t})\, dt = \int_0^\infty e^{-\lambda_2 t} \lambda_1 e^{-\lambda_1 t} dt$$

$$= \lambda_1 \int_0^\infty e^{-(\lambda_1 + \lambda_2)t} dt = \frac{\lambda_1}{\lambda_1 + \lambda_2}$$

NOTE

This result can be easily generalized to a system of n independent components with failure rates $\lambda_1, \lambda_2, \ldots, \lambda_n$. The probability that component j is the first one to fail is

$$P(\text{component } j \text{ fails first}) = \frac{\lambda_j}{\sum_{i=1}^{n} \lambda_i}$$

## 5.3 Parallel system

Suppose that in the system shown in Figure 5.1 the two components have the same cost, and their reliabilities are $R_1 = 0.7$, $R_2 = 0.95$, respectively. If it is permissible to add two components to the system, would it be preferable to a) replace component 1 by three components in parallel or b) to replace components 1 and 2 each by simple parallel systems?

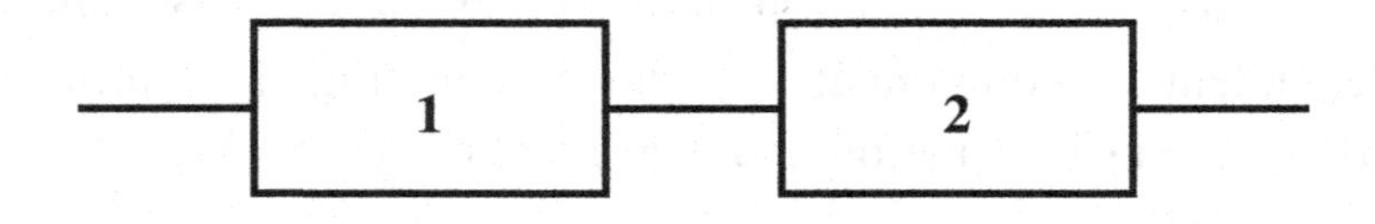


Figure 5.1. The system

### Solution

If component 1 is replaced by three components in parallel, then

$$R_a = [1-(1-R_1)^3] \cdot R_2 = 0.973 \times 0.95 = 0.92435 .$$

If each of the two components is replaced by a simple parallel system,

$$R_a = [1-(1-R_1)^2] \cdot [1-(1-R_2)^2] = 0.91 \times 0.9975 = 0.9077 .$$

In this problem the reliability $R_1$ is so low that even the reliability of a simple parallel system, $2R_1 - R_1^2$, is smaller than that of $R_2$. Thus replacing component 1 by three parallel components yields the higher reliability.

**5.4 Series-parallel system**

Suppose that in Figure 5.2, $R_{ai} = R_{bj} = e^{-\lambda t} \equiv R_*$, $i = 1, 2, 3, 4$ and $j = 1, 2$, and $R_c = 1$. Find the reliability $R$ of the system in the rare-event approximation.

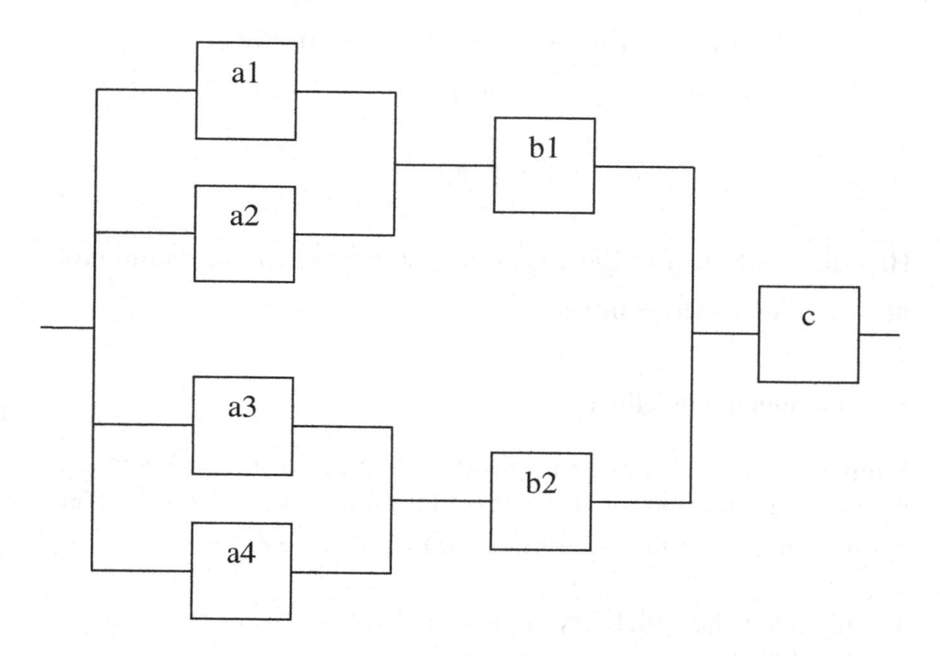


Figure 5.2. Reliability block diagram of a series-parallel configuration

**Solution**

$$R = R_*^2(2 - R_*)[2 - (2 - R_*)R_*^2]$$

Expanding the above term we get:

$$R = 4 \cdot R_*^2 - 2 \cdot R_*^3 - 4 \cdot R_*^4 - 4 \cdot R_*^5 - R_*^6$$

Then, we expand the exponential in a second-order Taylor series: $R_*^N = e^{-N\lambda t} \approx 1 - N\lambda t + \frac{1}{2}N^2(\lambda t)^2 - \ldots$ to obtain, for small $\lambda t$:

$$R \approx 4[1-2\lambda t+2(\lambda t)^2]-2[1-3\lambda t+\tfrac{9}{2}(\lambda t)^2]-4[1-4\lambda t+8(\lambda t)^2]$$
$$+4[1-5\lambda t+\tfrac{1}{2}25(\lambda t)^2]-1+6\lambda t-18(\lambda t)^2$$

$$R \approx (4-2-4+4-1)-(8-6-16+20-6)(\lambda t)$$
$$-(-8+9+32-50+18)(\lambda t)^2+\ldots$$

$$R \approx 1-(\lambda t)^2$$

Had the coefficient of the $(\lambda t)^2$ term also been zero, we would have needed to carry terms in $(\lambda t)^3$.

### 5.5 Common mode failure

Suppose that a unit has a design-life reliability of 0.95. Assume an exponential distribution for the failure times and the rare-event approximation for the reliability, $R(t) = e^{-\lambda t} \cong 1 - \lambda t$.

1. Estimate the reliability if two of these units are put in active parallel.
2. Consider now the possibility of common-mode failures (shocks which simultaneously fail both components). In this case, the failure rate $\lambda$ has two contributions from independent (I) and common-mode (C) failures

$$\lambda = \lambda_I + \lambda_C = (1-\beta)\lambda + \beta\lambda \text{ where } \beta = \frac{\lambda_C}{\lambda}$$

Estimate the maximum fraction $\beta$ of common failures that is acceptable if the parallel units in part 1 are to retain a system reliability of at least 0.99.

**Solution**

1. In general, $R(t) = 2e^{-\lambda t} - e^{-2\lambda t}$. In the rare-event approximation, $R(t) = 0.95 \approx 1 - \lambda t$ and, thus, $\lambda t = 0.05$.

$$R(t) \approx 1 - (\lambda t)^2 = 0.9975$$

2.
$$F = 1 - R = 0.01 = 1 - \left(2e^{-\lambda_I t} - e^{-2\lambda_I t}\right)e^{-\lambda_C t}$$
$$= 1 - \left(2 - e^{-(1-\beta)\lambda t}\right)e^{-\lambda t} \approx \beta\lambda t + \left(1 - 2\beta + \frac{\beta^2}{2}\right)(\lambda t)^2$$

Thus, with $\lambda t = 0.05$, we have:

$$0.00125\beta^2 + 0.045\beta - 0.0075 = 0$$

Therefore, $\beta = \dfrac{-0.045 \pm (2.0625 \times 10^{-3})^{1/2}}{0.0025}$

For $\beta$ to be positive, we must take the positive root. Therefore, $\beta \le 0.166$.

### 5.6 Active parallel system

In an active parallel system each unit has a failure rate of 0.002 $hr^{-1}$.

1. What is the *MTTF* of the system if there is no load sharing?
2. What is the *MTTF* of the system if the failure rate increases by 20% as a result of increased load?
3. What is the *MTTF* of the system if one simply (and conservatively) increased both unit failure rates by 20%?

**Solution**

1. $$MTTF = \frac{3}{2\lambda} = \frac{3}{2\times 0.002} = 750 \text{ hr}$$

2. In this case, $R(t) = 2e^{-\lambda^* t} + e^{-2\lambda t} - 2e^{-(\lambda+\lambda^*)t}$

$$MTTF = \int_0^\infty R(t)dt = \int_0^\infty \left[2e^{-\lambda^* t} + e^{-2\lambda t} - 2e^{-(\lambda+\lambda^*)t}\right]dt$$

or

$$MTTF = \frac{2}{\lambda^*} + \frac{1}{2\lambda} - \frac{2}{\lambda+\lambda^*}$$

Thus with $\lambda^* = 1.2\times 0.002 = 0.0024 \text{ hr}^{-1}$ we have

$$MTTF = \frac{2}{0.0024} + \frac{1}{2\times 0.002} - \frac{2}{0.0044} = 629 \text{ hrs}$$

3. $$MTTF = \frac{3}{2\lambda^*} = \frac{3}{2\times 0.0024} = 625 \text{ hr}$$

## 5.7 Shared load parallel system

In a "shared load parallel system," the partial components equally share the load, and, as a component fails, the surviving components must sustain an increased load. Thus, as successive components fail, the failure rates of the surviving components increase. An example of a shared parallel load configuration would be when bolts are used to hold a machine member; if one bolt breaks, the remainders must support the load.

Consider such a system with two components whose constant failure rates are defined as follows:

$\lambda_h$ = half-load failure rate

$\lambda_f$ = full-load failure rate

Find the time-dependent reliability of the system.

**Solution**

In solving this problem, the two states that should be taken into consideration are the following:

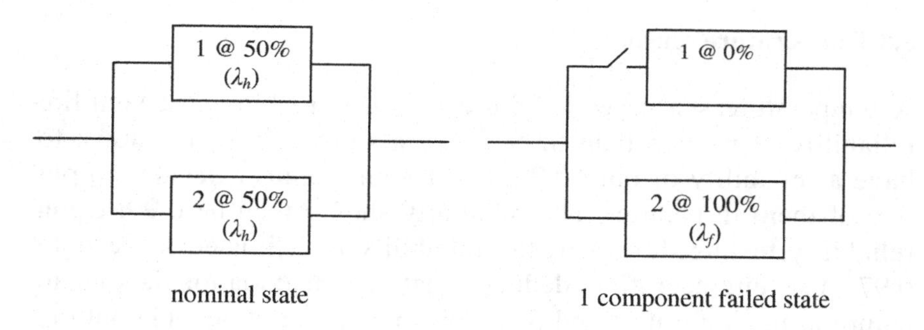


Reliability:

$$\begin{aligned} R(t) &= P\{\text{system successful up to } t\} \\ &= P\{\text{full load is adequately supported up to } t\} \\ &= P\{\text{1 and 2 survive up to t}\} \\ &+ \text{P}\{\text{1 fails at } \tau < t \text{ and 2 survives up to } \tau \text{with } \lambda_h \text{ and} \\ &\qquad \text{from } \tau \text{ to } t \text{ with } \lambda_f\} \\ &+ \text{P}\{\text{same conditions as before with 1 and 2 interchanged}\} \end{aligned}$$

$$R(t) = e^{-\lambda_h t} \times e^{-\lambda_h t} + \int_0^t \lambda_h e^{-\lambda_h \tau} e^{-\lambda_h \tau} e^{-\lambda_f (t-\tau)} d\tau + \int_0^t \lambda_h e^{-\lambda_h t} e^{-\lambda_h \tau} e^{-\lambda_f (t-\tau)} d\tau$$

$$= e^{-2\lambda_h t} + 2\lambda_h e^{-\lambda_f t} \int_0^t e^{-(2\lambda_h - \lambda_f)\tau} d\tau$$

$$= e^{-2\lambda_h t} + 2\lambda_h e^{-\lambda_f t} \times -\left(\frac{1}{2\lambda_h - \lambda_f}\right)\left[e^{-(2\lambda_h - \lambda_f)\tau}\right]\Big|_0^t$$

$$= e^{-2\lambda_h t} + \frac{2\lambda_h e^{-\lambda_f t}}{2\lambda_h - \lambda_f} \times \left[1 - e^{-(2\lambda_h - \lambda_f)t}\right]$$

$$= e^{-2\lambda_h t} + \frac{2\lambda_h}{2\lambda_h - \lambda_f} \times \left[e^{-\lambda_f t} - e^{-2\lambda_h t}\right]$$

## 5.8 Temperature sensor

A temperature sensor with failure rate $\lambda$ is to have a design-life reliability of no less than 0.98. Since a single sensor is known to have a reliability of only 0.90, the design engineer decides to put two of them in parallel. The reliability should then be 0.99. Upon reliability testing, however, the reliability is estimated to be only 0.97. The engineer first deduces that the degradation is due to common-mode failures and then considers two options: (1) putting a third sensor in parallel, and (2) reducing the probability of common-mode failures $\lambda_C$.

1. Assuming that the sensors have constant failure rates, find the value of $\beta = \lambda_C / \lambda$ that characterizes the common-mode failures.
2. Will adding a third sensor in parallel meet the reliability criterion if nothing is done about common-mode failures?
3. By how much must $\beta$ be reduced if the two sensors in parallel are to meet the criterion?

### Solution

1. Common-mode failures

If the design-life reliability of a sensor is

$$R_1 = e^{-\lambda T_d} = 0.9 \rightarrow \lambda T_d = 0.10536.$$

The reliability of two parallel sensors should then be:

$$R_2 = 2R_1 - R_1^2 = 0.99$$

However, the actual system reliability $\overline{R_2}$ for the two sensors in parallel is 0.97. The effect of a common mode failure mechanism is the same effect as putting in an additional component (with reliability $R_C$) in series with the parallel configuration. Thus, the reliability of this system ($\overline{R}_2$) becomes:

$$\overline{R}_2 = (2R_1 - R_1^2)R_C$$

If $\lambda_C$ is the common mode failure rate and $\lambda_I$ the rate of independent failures, we have:

$$\overline{R}_2 = (2e^{-\lambda_I T_d} + e^{-2\lambda_I T_d})e^{-\lambda_C T_d}$$

The total failure rate $\lambda$ of a single unit is:

$$\lambda = \lambda_I + \lambda_C$$

Introducing the factor $\beta$:

$$\beta = \frac{\lambda_C}{\lambda}$$

we may write

$$\overline{R}_2 = (2 - e^{-(1-\beta)\lambda T_d})e^{-\lambda T_d} = 0.97$$

Then, $\beta$ is found to be:

$$\beta = 1 + \frac{1}{\lambda T_d} \ln(2 - \overline{R}_2 e^{\lambda T_d}) = 0.2315$$

2. Effect of a third sensor

Let $X$ denote the event of the system failure and $X_i$ the event of $i$-th component failure. Thus for a system of 3 parallel components, we have

$$X = X_1 \cap X_2 \cap X_3$$

and the system reliability $R_3$, without any common mode failure, is:

$$R_3 = 1 - P(X_1 \cap X_2 \cap X)$$

If the failure are mutually independent, we may write:

$$R_3 = 1 - P(X_1)P(X_2)P(X_3)$$

The $P[X_i]$, $i = 1,2,3$ are the component failure probabilities.

$$P(X_i) = 1 - R_i$$

In this case, all the $R_i$ have the same value, $R_i = R_1$. Consequently,

$$R_3 = 1 - (1 - R_1)^3 = 1 - (1 - e^{-\lambda_I T_d})^3 = e^{-3\lambda_I T_d} - 3e^{-2\lambda_I T_d} + 3e^{-\lambda_I T_d}$$

If we consider the common mode failure, we can compute $\overline{R}_3$.

$$\overline{R}_3 = R_3 e^{-\lambda_C T_d}$$

We may expand the term to obtain:

$$\overline{R}_3 = [3 - 3e^{-(1-\beta)\lambda T_d} + e^{-2(1-\beta)\lambda T_d}]e^{-\lambda T_d} = 0.975$$

Thus the reliability criterion is not met.

3. Reduction in $\beta$ necessary to meet the $\lambda$=0.98 criterion

To meet the criterion with only two sensors in parallel, we must reduce $\beta$ enough so that the equation of $\beta$ in 1. is satisfied with $\overline{R}_2 = 0.98$. Thus:

$$\beta = 1 + \frac{1}{0.10536}\ln(2 - \frac{0.97}{0.9}) = 0.1165$$

Therefore, $\beta$ must be reduced at least to

$$1 - \frac{0.1165}{0.2315} \approx 50\%$$

### 5.9 A two engine plane

Consider a two engine plane in a one-out-of-two logic configuration. When both engines A and B are fully energized they share the total load and the failure time densities are $f_A(t)$ and $f_B(t)$. If either one fails, the survivor must carry the full load and its failure density becomes $g_A(t)$ or $g_B(t)$.

1. Derive an expression for the reliability of the system $R(t)$.
2. Find the reliability if

$$f_A(t) = f_B(t) = \lambda e^{-\lambda t}$$

$$g_A(t) = g_B(t) = k\lambda e^{-k\lambda t} \quad k > 1$$

**Solution**

1. Reliability expression

The reliability of the system is:

$R(t) = P$(system success at $t$)

$= P$(full load is adequately supported at $t$)

$= P$(component 1 and component 2

survive up to $t$ with failure densities $f_A(t)$ and $f_B(t)$)

$+P$(component 1 fails at any $\tau < t$ and component 2

survives up to $\tau$ with failure time density $f_B(t)$

and from $\tau$ to $t$ with failure time density $g_B(t)$)

$+P$(component 2 fails at any $\tau < t$ and component 1 survives

up to $\tau$ with failure time density $f_A(t)$

and from $\tau$ to $t$ with failure time density $g_A(t)$)

2. Reliability of the system

$$R(t) = [1 - F_A(t)] \cdot [1 - F_B(t)] + \int_0^t f_A(t')(1 - F_B(t'))(1 - G_B(t - t'))\,dt'$$

$$+ \int_0^t (1 - F_A(t'))(1 - G_A(t - t'))\,f_B(t')dt'$$

$$R(t) = e^{-\lambda t} e^{-\lambda t} + 2\int_0^t \lambda e^{-\lambda t^*} e^{-\lambda t^*} e^{-k\lambda(t - t^*)}\,dt^* = e^{-2\lambda t} + 2\lambda e^{-k\lambda t} \frac{\left[e^{-\lambda t^*(k-2)}\right]_0^t}{\lambda(k-2)}$$

$$R(t) = e^{-2\lambda t} + 2\lambda e^{-k\lambda t} \frac{e^{-\lambda t(k-2)} - 1}{\lambda(k-2)}$$

**5.10 One-out-of-two system**

Consider two components A and B in a one-out-of-two logic configuration. When both A and B are fully energized they share the total load and the failure densities are $f_A(t)$ and $f_B(t)$. If either one fails, the survivor must carry the full load and its failure density becomes $g_A(t)$ or $g_B(t)$. [A simple example would be a two-engine plane which, if one engine fails, can still keep flying, but the surviving engine now has to carry the full load.] Find the reliability $R(t)$ of the system if

$$f_A(t) = f_B(t) = \lambda e^{-\lambda t} \quad \text{and} \quad g_A(t) = g_B(t) = k\lambda e^{-\lambda k t}, \; k > 1$$

**Solution**

$$\begin{aligned}
R(t) &= P\{\text{system survives up to } t\} \\
&= P\{\text{neither component fails before } t\} \\
&+ P\{\text{one fails at some time } \tau < t, \\
&\text{the other one survives up to } \tau, \text{ with } f(t), \\
&\text{and from } \tau \text{ to } t \text{ with } g(t)\}
\end{aligned}$$

$$\begin{aligned}
R(t) &= e^{-2\lambda t} + 2\int_0^t \left(\lambda e^{-\lambda\tau} d\tau\right)\left(e^{-\lambda\tau}\right)\left(e^{-k\lambda(t-\tau)}\right) \\
&= e^{-2\lambda t} + 2\lambda e^{-k\lambda t}\int_0^t e^{-\lambda(2-k)\tau} d\tau \\
&= \frac{2e^{-k\lambda t} - ke^{-2\lambda t}}{2-k}
\end{aligned}$$

This is the solution for $k \neq 2$. If $k = 2$, we find that

$$R(t) = \left(1 + 2\lambda t\right)e^{-2\lambda t}.$$

### 5.11 Pressure vessel

A pressure vessel is equipped with six relief valves. Pressure transients can be controlled successfully by any three of these valves. If the probability that any one of these valves will fail to operate on demand is 0.04, what is the probability on demand that the relief valve system will fail to control a pressure transient? Assume that the failures are independent.

#### Solution

Let us define the unreliability of the generic component, $F = 1 - R$, as the demand failure probability, which is equal to 0.04 in our case. Using the rare-event approximation, we have, with $N = 6$ and $r = 3$:

$$F_{sys} = \sum_{k=r+1}^{N} \binom{N}{k} F^k (1-F)^{N-k}$$

Since $F$ is very small, $(1-F) \approx 1$ and only the term $F$ with smaller power $k = r+1$ gives a significant contribution to the sum. Thus,

$$F_{sys} = \binom{6}{4}(0.04)^4 = \frac{6!}{(2!)(4!)}(0.04)^4 = 15 \times 256 \times 10^{-8} = 0.38 \times 10^{-4}$$

### 5.12 r-out-of-N detection system

You are to design an $r$-out-of-$N$ detection system. The number of components, $N$, must be as small as possible to minimize cost. The fail-to-danger (the component is requested to detect an actually present danger but fails to do so) and the fail-safe (the system gives a false alarm in absence of danger) probabilities for the identical components are $q_d = 10^{-2}$ and $q_s = 10^{-2}$. Your design must meet the following criteria:

1. Probability of system fail-to-danger < $10^{-4}$.
2. Probability of system fail-safe < $10^{-2}$.
3. What values of $r$ and $N$ should be used?

**Solution**

Make a Table of unreliabilities (i.e. the failure probabilities) for fail-safe and fail-to-danger (in the Table 5.1 we have used the rare-event approximations)

Table 5.1.

| $r$-out-of-$N$ | $F_s$ | $F_d$ |
|---|---|---|
| 1/1 | $q_s = 10^{-2}$ | $q_d = 10^{-2}$ |
| 1/2 | $2q_s = 2\times10^{-2}$ | $q_d^2 = 10^{-4}$ |
| 2/2 | $q_s^2 = 10^{-4}$ | $2q_d = 2\times10^{-2}$ |
| 1/3 | $3q_s = 3\times10^{-2}$ | $q_d^3 = 10^{-6}$ |
| 2/3 | $3q_s^2 = 3\times10^{-4}$ | $3q_d^2 = 3\times10^{-4}$ |
| 3/3 | $q_s^3 = 10^{-6}$ | $3q_d = 3\times10^{-2}$ |
| 1/4 | $4q_s = 4\times10^{-2}$ | $q_d^4 = 10^{-8}$ |
| 2/4 | $6q_s^2 = 6\times10^{-4}$ | $4q_d^3 = 4\times10^{-6}$ |
| 3/4 | $4q_s^3 = 4\times10^{-6}$ | $6q_d^2 = 6\times10^{-4}$ |
| 4/4 | $q_s^4 = 10^{-8}$ | $4q_d = 4\times10^{-2}$ |

At least $N$ = four components are required to meet both criteria. They are met by a 2/4 system.

### 5.13 Cold standby system of two units

Consider a "cold" standby system of two units. The on-line unit has an *MTTF* of 2 years. When it fails, the standby unit comes on line and its *MTTF* is 3 years. Assume that each component has an exponential failure times distribution.

1. What is the probability density function of the system failure time?
2. What is the *MTTF* of the system?
3. Repeat 1 and 2, assuming that the two components are in parallel in a one-out-of-two configuration.

**Solution**

1. Probability density function

For the generic component with exponential distribution of failure times,

$$MTTF = \int_0^\infty tf(t)dt = \int_0^\infty \lambda te^{-\lambda t}dt = \frac{1}{\lambda}$$

Then, $\lambda_1 = \dfrac{1}{2yrs}$ and $\lambda_2 = \dfrac{1}{3yrs}$

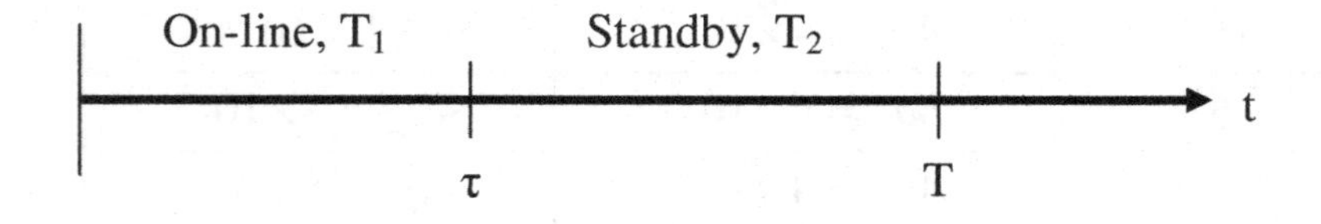


With reference to the above figure, $T_1$ and $T_2$ are independent random variables denoting the times when the on-line and standby units are operating, respectively.

The system failure time is also a random variable, $T = T_1 + T_2$.

The time $\tau$ is the random realization of the failure time of the on-line component:

$$f_{sys}(t) = \int_0^t \lambda_1 e^{-\lambda_1 \tau} \lambda_2 e^{-\lambda_2 (t-\tau)} d\tau = \lambda_1 \lambda_2 \int_0^t e^{-(\lambda_1 - \lambda_2)\tau} e^{-\lambda_2 t} d\tau$$

$$= \lambda_1 \lambda_2 e^{-\lambda_2 t} \int_0^t e^{-(\lambda_1 - \lambda_2)\tau} d\tau$$

$$= \frac{\lambda_1 \lambda_2}{\lambda_2 - \lambda_1} e^{-\lambda_2 t} \left( e^{-(\lambda_1 - \lambda_2)\tau} \right)_0^t$$

$$= \frac{\lambda_1 \lambda_2}{\lambda_2 - \lambda_1} \left( e^{-\lambda_1 t} - e^{-\lambda_2 t} \right)$$

$$= e^{-t/3\,yrs} - e^{-t/2\,yrs}$$

2. MTTF

$$MTTF = \int_0^\infty t f(t) dt = \int_0^\infty \left( te^{-t/3\,yrs} - te^{-t/2\,yrs} \right) dt$$

Let $u = \dfrac{t}{3yrs}$ and $\xi = \dfrac{t}{2yrs}$

$$MTTF = (3yrs)^2 \left[ -ue^{-u} - e^{-u} \right]_0^\infty - (2yrs)^2 \left[ -\xi e^{-\xi} - e^{-\xi} \right]_0^\infty$$

$$= (3yrs)^2 \left( \frac{1}{yr} \right) - (2yrs)^2 \left( \frac{1}{yr} \right) = 5 \text{ yrs}$$

3. Parallel system

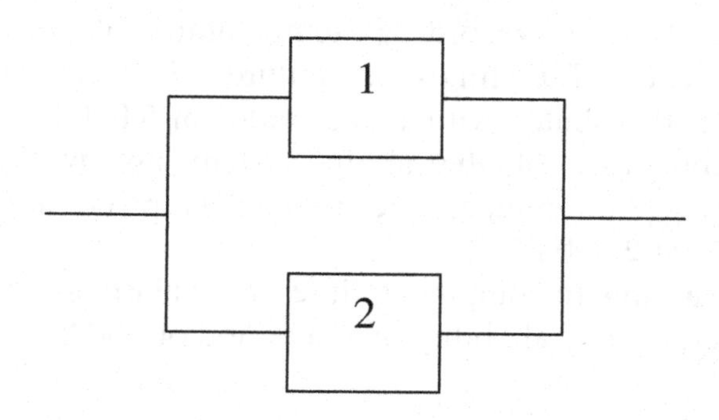

$$P_{failure} = P\{T \le t\} = P[\{T_1 \le t\} \cap \{T_2 \le t\}]$$
$$= F[T_1 \le t] \cdot F[T_2 \le t] = F_{sys}[T \le t]$$
$$= (1 - e^{-\lambda_1 t})(1 - e^{-\lambda_2 t}) = 1 - e^{-\lambda_1 t} - e^{-\lambda_2 t} + e^{-(\lambda_1 + \lambda_2)t}$$

Then:

$$f_{sys}(t) = \frac{dF_{sys}}{dt} = \lambda_1 e^{-\lambda_1 t} + \lambda_2 e^{-\lambda_2 t} - (\lambda_1 + \lambda_2) e^{-(\lambda_1 + \lambda_2)t}$$
$$= \frac{1}{2} e^{-\frac{t}{2}} + \frac{1}{3} e^{-\frac{t}{3}} - \frac{5}{6} e^{-\frac{5t}{6}}$$

$$MTTF = \int_0^\infty t\left(f_{sys}(t)\right)dt = \int_0^\infty R_{sys}(t)dt = \int_0^\infty \left(1 - F_{sys}(t)\right)dt$$
$$= \int_0^\infty \left(e^{-\lambda_1 t} + e^{-\lambda_2 t} - e^{-(\lambda_1 + \lambda_2)t}\right)dt$$
$$= \frac{1}{\lambda_1} + \frac{1}{\lambda_2} - \frac{1}{\lambda_1 + \lambda_2} = 2 + 3 - \frac{6}{5} = \frac{19}{5} = 3.8 \text{ yrs}$$

## 5.14 Temperature sensing elements

Three nominally identical temperature sensing elements are connected to nominally the same point on a process plant. An alarm is designed to be given if any two or more of these temperature sensors record a temperature above a certain prescribed level. The times to failure of each element are exponentially distributed with a mean value of 5,000 h. What is:

1. The probability of the alarm system not working, if an excessive plant temperature rise takes place at 500 h, or secondly at 2,000 h?
2. The mean time to complete failure of the alarm system?
3. The average unavailability over a period of 500 h?

**Solution**

For each sensor, $p = 1 - e^{-\lambda t}$ with $\lambda^{-1} = MTTF = 5000$ hrs.

1. Probability of the alarm system not working

Let the top event be:

$T =$ 'the alarm system is not working'.

The fault tree is:

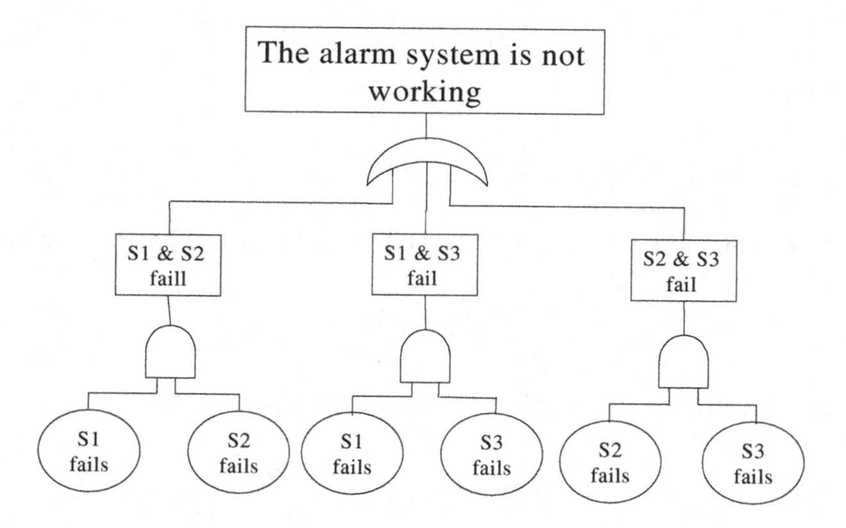


The structure function $X_T \equiv \Phi(X_1, X_2, X_3)$ is:

$$X_T = 1 - (1 - X_1X_2) \cdot (1 - X_1X_3) \cdot (1 - X_2X_3)$$
$$= X_1X_2 + X_1X_3 + X_2X_3 - 2X_1X_2X_3$$

If an excessive plant temperature rise takes place at time t, the instantaneous probability, $q(t)$, of the alarm system not working at time t is:

$$q(t) = E[\Phi(X_1, X_2, X_3)]$$
$$= E[X_1] \cdot E[X_2] + E[X_1] \cdot E[X_3] + E[X_2] \cdot E[X_3]$$
$$-2 \cdot E[X_1] \cdot E[X_2] \cdot E[X_3]$$

The times to failure of each element are exponentially distributed with a mean value of 5000 h, thus:

$$\lambda = \frac{1}{5000} h^{-1}$$

and the probability, $p_i(t)$ that the i-th sensor failed before t is

$$p_i(t) = p(t) = 1 - e^{-\lambda t}$$

Then,

$$q(t) = 3p^2(t) - 2p^3(t) = 1 - 3e^{-2\lambda t} + 2e^{-3\lambda t}$$

If $t_1 = 500$ hrs we have $Q(t_1) = 0.025$ and if $t_2 = 2000$ hrs we have

$$Q(t_2) = 0.25.$$

2. Mean time to complete failure of the alarm system

$$R(t) = 1 - q(t) = 3e^{-2\lambda t} - 2e^{-3\lambda t}$$

$$MTTF = \int_0^{+\infty} R(t)dt = \left| \frac{-3e^{-2\lambda t}}{2\lambda} + \frac{2e^{-3\lambda t}}{3\lambda} \right|_0^{+\infty} = \frac{5}{6\lambda} = 4166.7$$

3. Average unavailability over a period f 500 h

$$U(t) = \frac{\int_0^T q(t)dt}{T} = \frac{\left| t + \frac{3e^{-2\lambda t}}{2\lambda} - \frac{2e^{-3\lambda t}}{3\lambda} \right|_0^T}{T}$$

$$U(T = 500) = 8,84 \cdot 10^{-3}$$

# Chapter 6

# Availability and maintainability

## 6.1 Compressor

A compressor is designed for $T_d = 5$ years of operation. There are two significant contributions to the failure. The first is due to wear (W) of the thrust bearing and is described by a Weibull distribution

$$f(t) = \frac{m}{\vartheta}\left(\frac{t}{\vartheta}\right)^{m-1} e^{\left[-\left(\frac{t}{\vartheta}\right)^{m}\right]}$$

with $\vartheta = 7.5$ year and $m = 2.5$. The second, which includes all other causes (O) is described by a constant failure rate of $\lambda_0 = 0.013$ (year)$^{-1}$.

1. What is the reliability if no preventive maintenance is performed over the 5-year design life?
2. If the reliability of the 5-year design life is to be increased to at least 0.9 by periodically replacing the thrust bearing, how frequently must it be replaced?
3. Suppose that the probability of fault bearing replacement causing failure of the compressor is $p = 0.02$. What will the design-life reliability be with the replacement program decided in 2)?

**Solution**

1. Reliability with no preventive maintenance

The reliability of a system with two different failure modes (W and O) is:

$$R(t) = P(X_w \cap X_o)$$

where $X_W$ is the event in which the wear failure mode represented by the Weibull distribution does not occur before time $t$ and $X_O$ is the event in which the failure mode with constant failure rate does not occur before time $t$. Since the modes are independent, we may write the system reliability as the product of the mode survival probabilities:

$$R(t) = P(X_w)P(X_o) = R_w(t)R_o(t)$$

Note that $R_w(t)$ is the reliability if only the thrust bearing wear is considered and $R_o(t)$ is the reliability if only the constant failure rate is considered.

Thus,

$$R(T_d) = R_w(T_d) \cdot R_o(T_d) = e^{-\left(\frac{T_d}{\vartheta}\right)^m} \cdot e^{-\lambda_o T_d} = 0.6957 \cdot 0.9371 = 0.6519$$

2. Frequency of replacement to achieve 0.9 reliability

Preventive maintenance:

Suppose that we divide the design life into $N$ equal intervals; the time interval, at which maintenance is carried out is then $T = T_d / N$. Correspondingly, $T_d = NT$.

If we perform maintenance at $T$, restoring the system to an as-good-as-new condition, the system at $t > T$ has no memory of accumulated wear effects at times before $T$. Thus, in the interval $T < t \leq 2T$, the reliability is the product of the probability $R(T)$ that the system survived to $T$, and the probability $R(t-T)$ that a system as good as new at $T$ will survive for a time $t$-$T$ without failure:

$$R(t) = R(T)R(t-T) \quad , \quad T \le t < 2T$$

The same arguments may be used repeatedly to obtain the general expression

$$R(t) = R(T)^N R(t-NT) \quad , \quad NT \le t < (N+1)T \quad , \quad N = 0,1,2,...$$

Since the system reliability at time $T_d$ is the product of the time survival probabilities:

$$R(T_d) = R_w(T_d)R_o(T_d)$$

We can calculate separately the two reliability contributions. For bearing replacement at time $T_d = NT$, we have

$$R_W(T_d) = e^{-\left[N\left(\frac{T_d}{N\vartheta}\right)^m\right]} = e^{\left[-N^{1-m}\left(\frac{T_d}{\vartheta}\right)^m\right]}$$

and for the constant failure rate we have:

$$R_o(T_d) = e^{-\lambda_o NT} = e^{-\lambda_o T_d} = 0.9371$$

which is exactly the same reliability of the system if no preventive maintenance is performed. Thus, as expected, in case of failures occurring with a constant failure rate $\lambda_o$, preventive maintenance has no effect.

For the criterion of 0.9 reliability to be met, we must have:

$$R_w(T_d) = \frac{R(T_d)}{R_o(T_d)} \ge \frac{0.9}{0.9371} \quad \Rightarrow \quad R_w(T_d) \ge 0.9604$$

With $\left(\frac{T_d}{\vartheta}\right)^m = \left(\frac{5}{7.5}\right)^{2.5} = 0.36289$, we calculate

$$R_w(T_d) = e^{0.36289 N^{-1.5}}$$

Table 6.1.

| $N$ | 1 | 2 | 3 | 4 | 5 |
|---|---|---|---|---|---|
| $R_w(T_d)$ | 0.696 | 0.880 | 0.933 | 0.956 | 0.968 |

Thus the criterion is met for $N$=5, and the time interval for bearing replacement is $T = T_d / N = 1$ year.

3. Reliability in case of replacement-driven failures

At the end of the design life ($T_d = 5$ years) maintenance will have been performed four times. In case of perfect maintenance we have:

$$R(T_d) = R_o R_w = 0.937 \cdot 0.968 = 0.907$$

whereas, with imperfect maintenance,

$$R(T_d) = R_o R_w (1-p)^4 = 0.907 \cdot 0.98^4 = 0.836$$

## 6.2 Sequential and staggered maintenance scheme of a one-out-of-two system

Consider a one-out-of two system of identical components with constant failure rate $\lambda$. The testing and repair of each component last for $\tau_r$ hours.

1. In the sequential maintenance scheme, the two components are tested one after the other, $\tau$ being the time between the end of the previous maintenance of the second component and the

beginning of the next maintenance of the first one (in other words, every $\tau$ hours we test both components in sequence). Find the average unavailability of the system.

2. In the staggered maintenance scheme, the first component maintenance starts at $k\tau$, $k < 1$, where $\tau$ is the time interval between the end of the previous maintenance of the second component and the beginning of the next maintenance of the same second component. Find the average unavailability of the system.

**Solution**

To solve this problem with regards to the two different maintenance schemes, we use the following general procedure:

1. Calculate instantaneous unavailability and average system downtime $< D_i >$ in every subinterval $T_i$ within a period $T$
2. Compute the average unavailability:

$$\overline{q} = \frac{\int_0^T q(t)dt}{T} = \frac{\sum_i < D_i >}{T}$$

For academic purposes, we start considering a single-component system undergoing periodic maintenance. Note that the system has a periodic behaviour because the maintenance restores the component to an as-good-as-new condition. With the maintenance scheme of the following Figure the period is $T = \tau + \tau_r$.

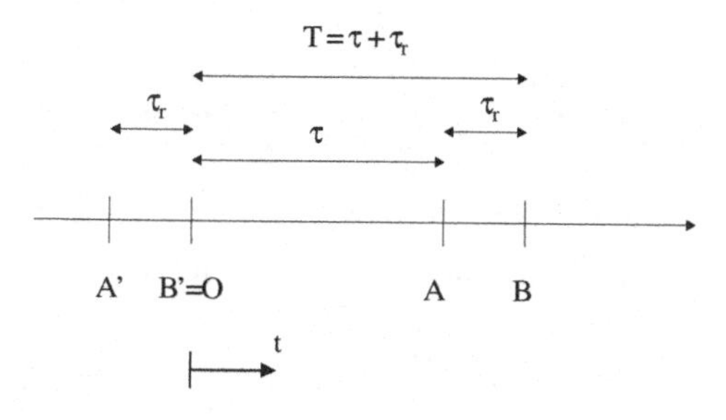

In the two time subintervals OA and AB the system has a different physical behaviour:

OA:

The component is unattended (no repair is allowed), thus the instantaneous unavailability, q(t) of the component will be:

$$q(t) = F(t)$$

and the average time the system is down, $< D_{\underline{OA}} >$, will be:

$$< D_{\underline{OA}} >= \int_0^\tau q(t)\,dt = \int_0^\tau \left(1 - e^{-\lambda t}\right) dt \approx \frac{\lambda \tau^2}{2}$$

AB

The component is always under repair, thus:

$$q(t) = 1$$

$$< D_{\underline{AB}} >= \int_\tau^{\tau+\tau_r} q(t)dt = \tau_r$$

The mean downtime over the period T is:

$$\bar{q} = \frac{\int_0^T q(t)dt}{T} = \frac{\int_0^\tau F(t)dt + \tau_r}{\tau + \tau_r} = \frac{\frac{\lambda \tau^2}{2} + \tau_r}{\tau + \tau_r} \approx \frac{\lambda \tau}{2} + \frac{\tau_r}{\tau}$$

1. Sequential maintenance scheme

We now consider a one-out-of-two system, i.e. a parallel system of two units. The following Figure represents the sequential maintenance scheme:

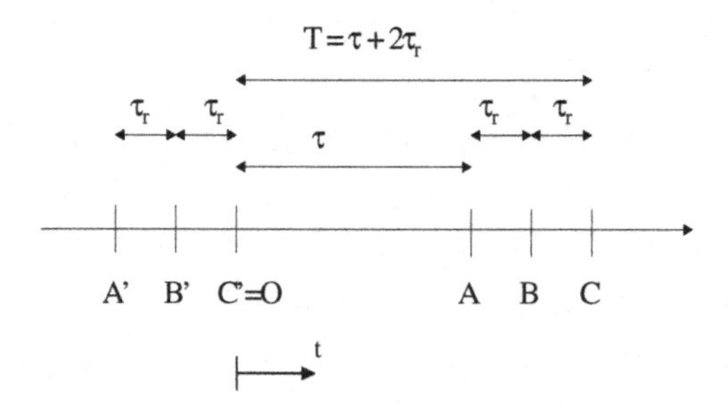

We split the time interval $T=\tau+2\tau_r$ in the 3 subintervals: OA, AB, BC, and we compute the average time the system is down in each subinterval.

OA:

Both components unattended, the first one up since B', the second since O.

$$q(t)=(1-e^{-\lambda(t+\tau_r)})\cdot(1-e^{-\lambda t})\approx\lambda(t+\tau_r)\lambda t$$

$$<D_{OA}>=\int_0^t q(t)dt=\lambda^2\left(\frac{\tau^3}{3}+\frac{\tau_r\tau^2}{2}\right)\approx\frac{\lambda^2\tau^3}{3}$$

AB:

First component is under maintenance, the second one is unattended since O.

$$q(t)=1-e^{-\lambda t}\approx\lambda t$$

$$<D_{AB}>=\int_\tau^{\tau+\tau_r} q(t)dt=\text{ (with } t'=t-\tau\text{ )}$$

$$=\int_0^{\tau_r}\left(1-e^{-\lambda(t'+\tau)}\right)dt'\approx\int_0^{\tau_r}\lambda(t'+\tau_r)dt'=\lambda\tau\tau_r+\frac{\lambda\tau_r^{\,2}}{2}\approx\lambda\tau\tau_r$$

BC:

First component up since B, second component under maintenance.

$$q(t)=1-e^{-\lambda(t-\tau_r-\tau)}\approx\lambda(t-\tau_r-\tau)$$

$$< D_{BC} >= \int_{\tau+\tau_r}^{\tau+2\tau_r} q(t)dt = \quad \text{with } t' = t - (\tau + \tau_r)$$

$$= \int_0^{\tau_r} \left(1 - e^{-\lambda t'}\right) dt' \approx \int_0^{\tau_r} \lambda t' \, dt' \approx \frac{\lambda \tau_r^{\ 2}}{2} \approx 0$$

The mean downtime is:

$$\bar{q} = \frac{\frac{\lambda^2\tau^3}{3} + \lambda\tau\tau_r + \frac{\lambda\tau_r^{\ 2}}{2}}{\tau + 2\tau_r} \approx \frac{\lambda^2\tau^2}{3} + \lambda\tau_r + \frac{\lambda\tau_r^{\ 2}}{2} \approx \frac{\lambda^2\tau^2}{3} + \lambda\tau_r$$

We can see that the only two meaningful terms are due to the failure of both component during OA and to the failure of the second component when the first is under repair during AB.

2. Staggered maintenance scheme

Consider the staggered maintenance scheme of the following Figure:

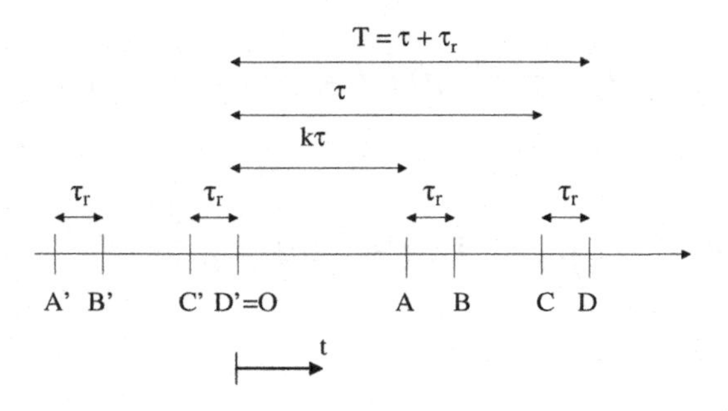

<u>OA</u>:
Both components are up since the end of their respective last test (D' and B').

$$q(t)=\left(1-e^{-\lambda(t+\tau-k\tau)}\right)\cdot\left(1-e^{-\lambda t}\right)\approx\lambda\left(t+(1-k)\tau\right)\lambda t$$

$$<D_{OA}>=\int_0^{k\tau}q(t)dt=\lambda^2\left(\frac{k^3\tau^3}{3}+(1-k)\frac{k^2\tau^3}{2}\right)\approx\frac{\lambda^2(3k^2-k^3)\tau^3}{6}$$

<u>AB</u>:
First component is under maintenance, second component up since O, end of its last test

$$q(t)=1-e^{-\lambda t}\approx\lambda t$$

$$<D_{AB}>=\int_{k\tau}^{k\tau+\tau_r}q(t)dt$$

Substituting with $t'=t-k\tau$ we have:

$$<D_{AB}>=\int_{k\tau}^{k\tau+\tau_r}q(t)dt=\int_0^{\tau_r}\lambda(t'+k\tau)dt'=\lambda k\tau\tau_r+\frac{\lambda\tau_r^{\ 2}}{2}\approx\lambda k\tau\tau_r$$

<u>BC</u>:
Second component up since O, first component up since B, end of its last test.

$$q(t)=\left(1-e^{-\lambda(t-\tau_r-k\tau)}\right)\cdot\left(1-e^{-\lambda t}\right)\approx\lambda\left(t-\tau_r-k\tau\right)\lambda t$$

$$<D_{BC}>=\int_{k\tau+\tau_r}^{\tau}\lambda(t-\tau_r-k\tau)\lambda t dt$$

Substituting with $t' = t - k\tau - \tau_r$ we have:

$$< D_{BC} >= \int_0^{\tau-\tau_r-k\tau} \lambda(t'+\tau_r+k\tau)\lambda t' dt' \approx \frac{\lambda^2(k^3-3k+2)\tau^3}{6}$$

CD:
Second component under maintenance, first component up since B, end of its last test

$$q(t) = \left(1 - e^{-\lambda(t-\tau_r-k\tau)}\right)$$

$$< D_{BC} >= \int_{\tau}^{\tau+\tau_r} \lambda(t-\tau_r-k\tau)dt = \text{ (with t'=t-}\tau\text{)}$$

Substituting with $t' = t - \tau$ we have:

$$< D_{BC} >= \int_0^{\tau_r} \lambda\left(t'-\tau_r+(1-k)\tau\right)dt' \approx \frac{\lambda\tau_r^{\ 2}}{2}$$
$$+(1-k)\lambda\tau\tau_r - \lambda\tau_r^2 \approx (1-k)\lambda\tau\tau_r$$

The mean downtime over $T$ is:

$$\bar{q} = \frac{\frac{\lambda^2(3k^2-k^3)\tau^3}{6} + \lambda k\tau\tau_r + \frac{\lambda^2(k^3-3k+2)\tau^3}{6} + (1-k)\lambda\tau\tau_r}{\tau+\tau_r} \approx$$
$$\approx \frac{\lambda^2(3k^2-3k+2)\tau^2}{6} + \lambda\tau_r$$

Chapter 7

# Fault tree analysis

## 7.1 Coolant supply system

Draw a fault tree for the coolant supply system pictured in Figure 7.1. Here the top event is loss of minimum flow to a heat exchanger.

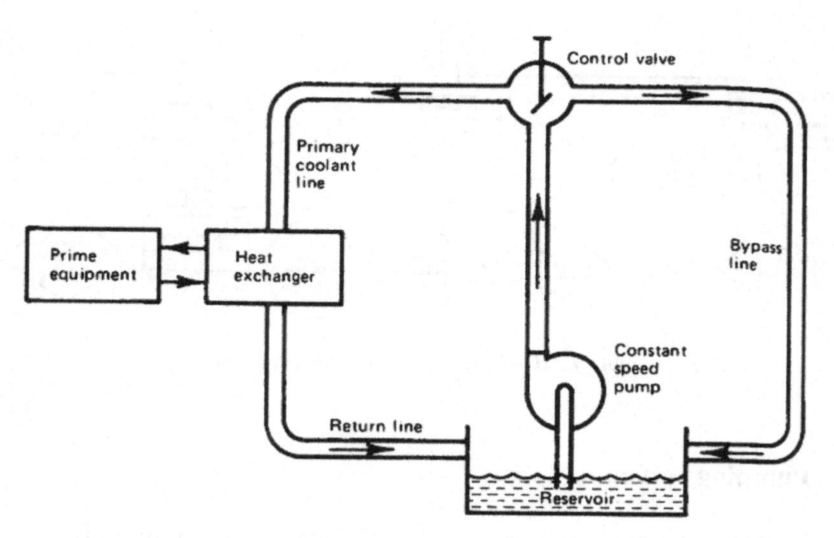


Figure 7.1. Coolant supply system

## Solution

The fault tree is shown in Figure 7.2. Not all of the faults at the bottom of the tree are primary failures. Thus it may be desirable to develop some of the faults, such as loss of the pump inlet supply, further. Conversely, the faults may be considered too insignificant to be traced further, or data may be available even though they are not primary failures.

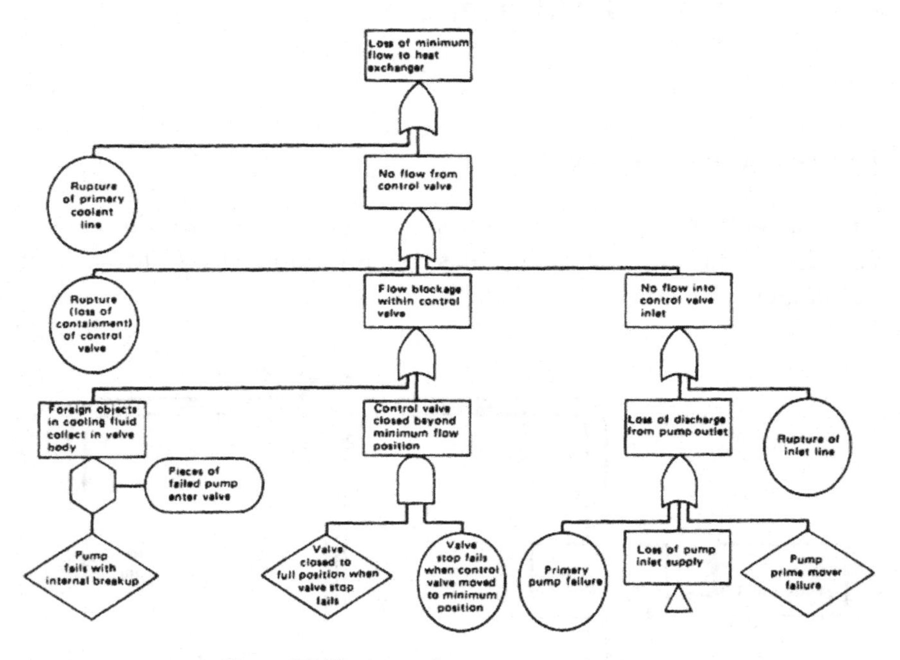


Figure 7.2. Fault tree for coolant supply system

## 7.2 Pumping system

In the pumping system shown in the Figure 7.3, the tank is filled in 10 min. and empties in 50 min.; thus, the cycle time is 1 hr. After the switch is closed, the timer is set to open the contacts in 10 min. If the mechanisms fail, then the alarm horn sounds and the operator opens the switch to prevent a tank rupture due to overfilling.

Consider the operator as a component: a primary failure would mean that the operator functioning within the design envelope fails to push the panic button when the alarm sounds; the secondary operator failure could be, for example, that the operator has fainted due to the smoke of a fire, when the alarm sounded. For all components (including the operator) assume:

$$P(\text{primary failure}) = 0.1\,;$$
$$P(\text{secondary failure}) = 0.05\,.$$

For the top event *tank rupture*:

1. Draw the fault tree;
2. Compute the probability of the top event;

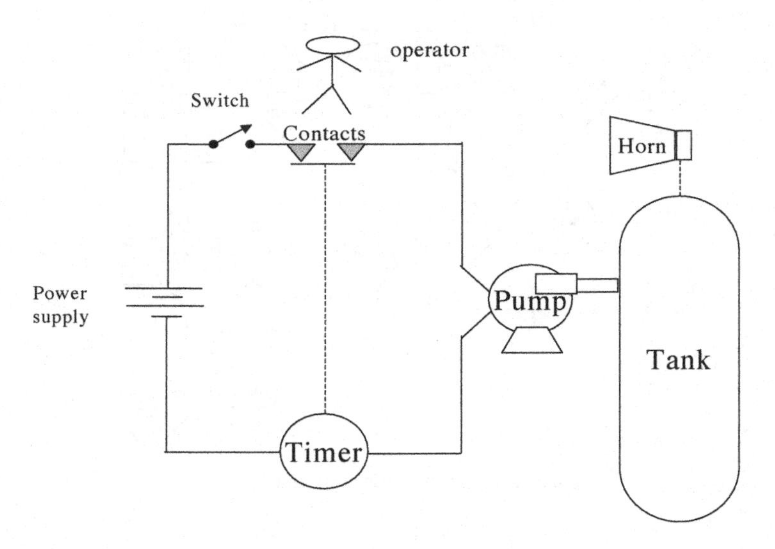


Figure 7.3. Pumping system

**Solution**

1. Fault tree is presented in the Figure 7.4:

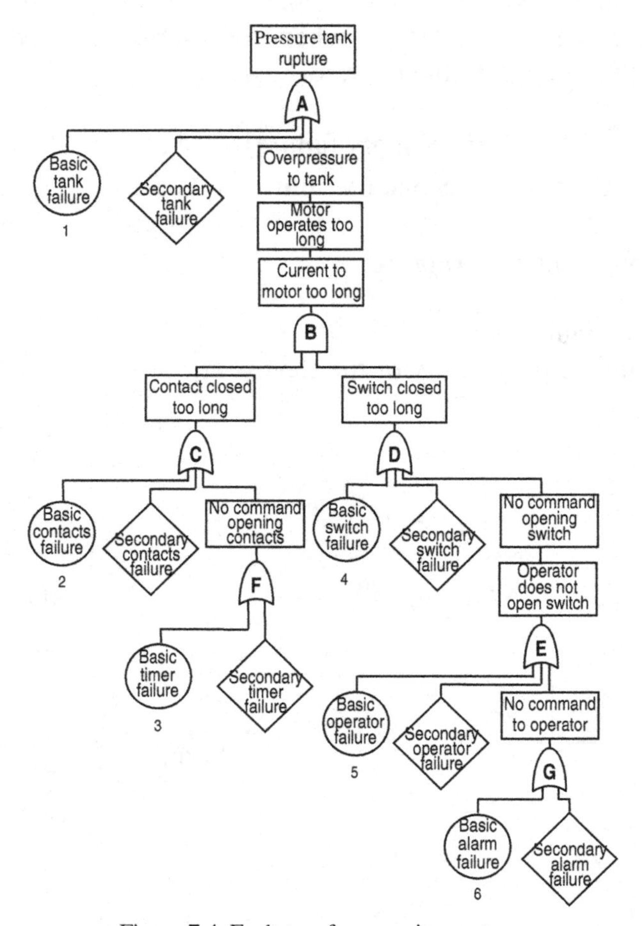


Figure 7.4. Fault tree for pumping system

2. Multiplying the probabilities of the events through the fault tree, as indicated in the Figure, one readily obtains:

$$P(X_T = 1) = 0.285$$

## 7.3 Structure function

Reduce the structure function $\Phi(\underline{X})$ to find the minimal cut sets and construct the corresponding fault tree with an AND gate.

$$\Phi(\underline{X}) = X_1X_2 + X_1X_3 + X_4 - X_1X_2X_3$$
$$-X_1X_3X_4 - X_1X_2X_4 + X_1X_2X_3X_4$$

**Solution**

The structure function

$$\Phi(\underline{X}) = X_1X_2 + X_1X_3 + X_4 - X_1X_2X_3$$
$$-X_1X_3X_4 - X_1X_2X_4 + X_1X_2X_3X_4$$

has to be manipulated to get the minimal cut sets. We firstly focus on the term $X_4$. It has to be a minimal cut set because it appears alone in the structure function with the sign "+" and thus cannot be the result of a product of other minimal cut sets. The step for the successive groupings of the terms down to the minimal cut sets are:

$$\begin{aligned}
\Phi(\underline{X}) &= X_1X_2 + X_1X_3 + X_4 - X_1X_2X_3 - X_1X_3X_4 \\
&-X_1X_2X_4 + X_1X_2X_3X_4 \\
&= 1-(1 - X_1X_3 - X_1X_2 + X_1X_2X_3 - X_4 + X_1X_3X_4 \\
&+X_1X_2X_4 - X_1X_2X_3X_4) \\
&= 1-\begin{bmatrix} 1 - X_1X_3 - X_1X_2 + X_1X_2X_3 \\ -X_4 \cdot (1 - X_1X_3 - X_1X_2 + X_1X_2X_3) \end{bmatrix} \\
&= 1-(1-X_4)\cdot\left[1 - X_1X_3 - X_1X_2(1 - X_1X_3)\right] \\
&= 1-(1-X_4)\cdot(1-X_1X_2)\cdot(1-X_1X_3)
\end{aligned}$$

The corresponding fault tree is:

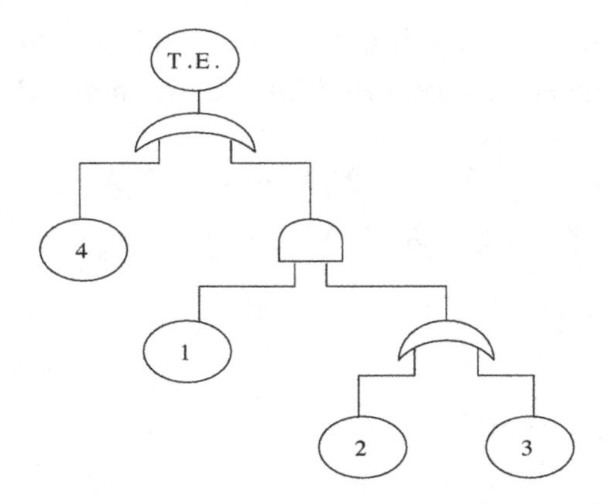


Figure 7.5. Fault tree

### 7.4 Fault tree and structure function

Construct the fault tree for the failure of the system in Figure, write the system structure function and reduce it to obtain the minimal cut sets.

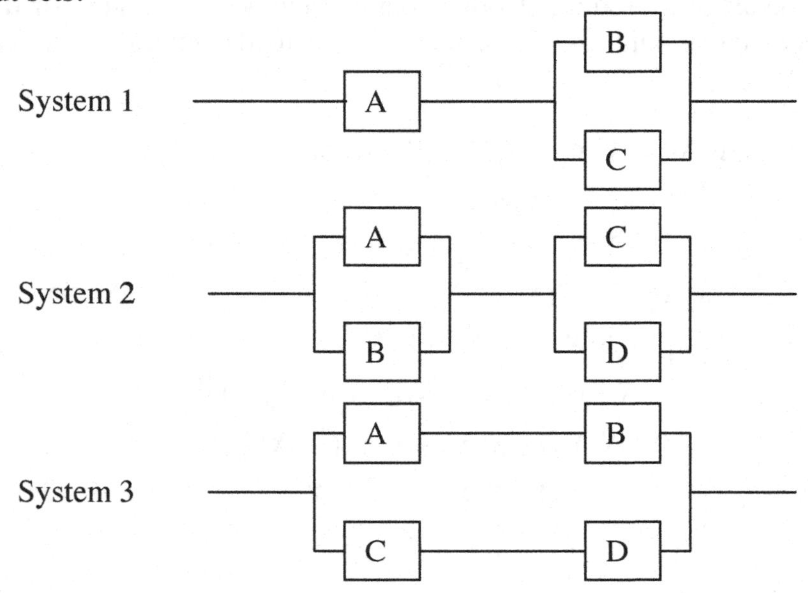

**Solution**

**System 1:** The system fault tree:

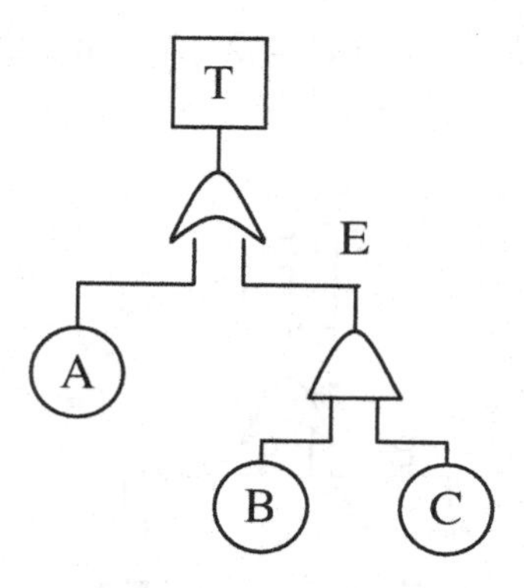


Figure 7.6. Fault tree

For the tree components' failures A, B, C, for the intermediate failure event E and for the top event T of system failure, we introduce the logic variables $X_A$, $X_B$, $X_C$, $X_E$, $X_T$, denoting whether the corresponding event has occurred or not. The variables $X_A$, $X_B$, $X_C$, $X_E$, $X_T$ assume the value 1 (true) if the corresponding event has occurred and the value 0 (false) if the event has not occurred.
By descending the fault tree with the rules of Boolean logic, we obtain the system structure function by writing $X_T$ as a function of the three variables $X_A$, $X_B$, $X_C$, thus:

$$X_T = 1-(1-X_A)(1-X_E)$$
$$X_E = X_B X_C$$
$$X_T = 1-(1-X_A)(1-X_B X_C)$$
$$X_T = X_A + X_B X_C - X_A X_B X_C$$

In principle, the equations above can be processed with the rules of Boolean logic to obtain an expression of the system structure function as a logic OR linking the system minimal cut sets. Actually, this is the case of the form of the indicator variable $X_T$, in which this is expressed as a logic OR among the events $X_A$ and $X_B X_C$. Thus, $M_1 = X_A$ and $M_2 = X_B X_C$ are the system minimal cut sets.

**System 2**:

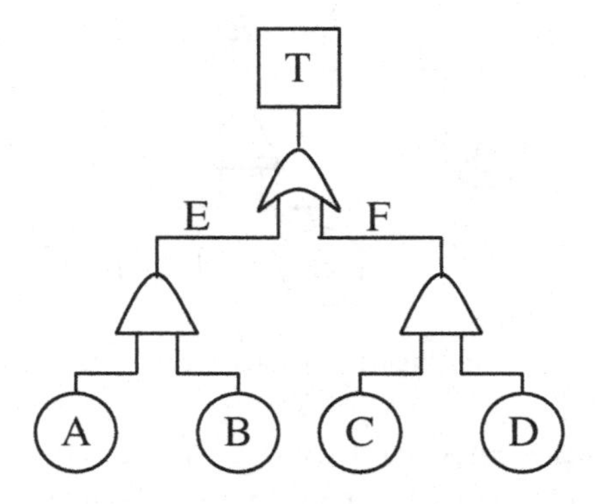


Figure 7.7. Fault tree

We introduce the indicator variables $X_T$, $X_A$, $X_B$, $X_C$, $X_D$, $X_E$, $X_F$ for the events T, A, B, C, D, E, F, G. The meaning has already been illustrated with reference to System 1.

$$X_T = 1-(1-X_E)(1-X_F)$$
$$X_E = X_A X_B$$
$$X_F = X_C X_D$$
$$X_T = 1-(1-X_A X_B)(1-X_C X_D)$$
$$X_T = X_A X_B + X_C X_D - X_A X_B X_C X_D$$

Analogous considerations to those made for the last two equations of system 1, stand for the last two equations for the present system. Therefore, the minimal cut-sets are readily identified as $M_1 = X_A X_B$ and $M_2 = X_C X_D$.

**System 3**:

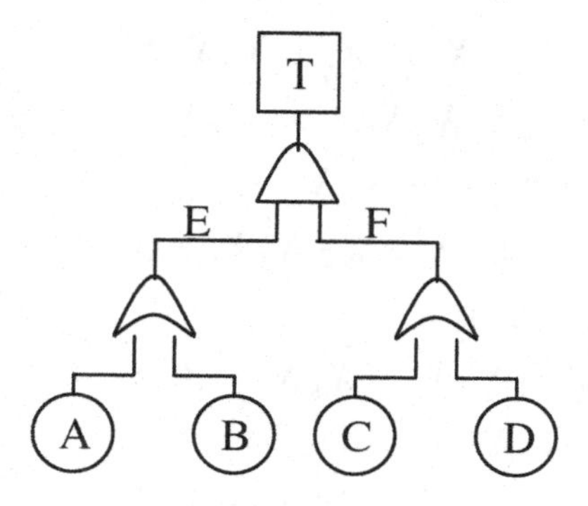


Figure 7.8. Fault tree

The system structure function, which expresses the indicator variable $X_T$ as a function of the basic failure events $X_A$, $X_B$, $X_C$, $X_D$, is obtained as follows.

$$X_T = X_E X_F$$

$$X_E = 1-(1-X_A)(1-X_B)$$
$$X_F = 1-(1-X_C)(1-X_D)$$

$$X_T = X_A X_C + X_A X_D - X_A X_C X_D + X_B X_C$$
$$+X_B X_D - X_B X_C X_D - X_A X_B X_C - X_A X_B X_D + X_A X_B X_C X_D$$

Now, the above expression has to be manipulated to get the minimal cut sets. The steps for the successive grouping of the terms down to the minimal cut sets are

$$X_T = X_A X_C + X_A X_D - X_A X_C X_D + X_B X_C$$
$$+X_B X_D - X_B X_C X_D - X_A X_B X_C - X_A X_B X_D + X_A X_B X_C X_D$$
$$= 1 - 1 + X_A X_C + X_A X_D - X_A X_C X_D + X_B X_C$$
$$+X_B X_D - X_B X_C X_D - X_A X_B X_C - X_A X_B X_D + X_A X_B X_C X_D$$

$$
\begin{aligned}
&= 1 - 1 + X_A X_C + X_A X_D - X_A X_C X_D + X_B X_C \\
&+ X_B X_D - X_B X_C X_D - X_A X_B X_C - X_A X_B X_D \\
&+ X_A X_B X_C X_D + X_A X_B X_C X_D - X_A X_B X_C X_D \\
&= 1 - [(1 - X_A X_D - X_B X_C - X_B X_D + X_B X_C X_D \\
&X_A X_B X_D) - X_A X_C (1 - X_A X_D - X_B X_C - X_B X_D \\
&X_B X_C X_D + X_A X_B X_D)] \\
&= 1 - (1 - X_A X_C)(1 - X_A X_D - X_B X_C - X_B X_D \\
&+ X_B X_C X_D + X_A X_B X_D) \\
&= 1 - (1 - X_A X_C)(1 - X_A X_D - X_B X_C - X_B X_D \\
&+ X_B X_C X_D + X_A X_B X_D + X_A X_B X_C X_D - X_A X_B X_C X_D) \\
&= 1 - (1 - X_A X_C)[(1 - X_B X_C - X_B X_D \\
&+ X_B X_C X_D) - X_A X_D (1 - X_B X_C - X_B X_D + X_B X_C X_D)] \\
&= 1 - (1 - X_A X_C)(1 - X_A X_D)(1 - X_B X_C - X_B X_D + X_B X_C X_D) \\
&= 1 - (1 - X_A X_C)(1 - X_A X_D)[(1 - X_B X_C) - X_B X_D (1 - X_B X_C)] \\
&= 1 - (1 - X_A X_C)(1 - X_A X_D)(1 - X_B X_C)(1 - X_B X_D)
\end{aligned}
$$

By so doing, four minimal cut sets are identified:

$$
\begin{aligned}
M_1 &= X_A X_C \\
M_2 &= X_A X_D \\
M_3 &= X_B X_C \\
M_4 &= X_B X_D
\end{aligned}
$$

### 7.5 Network system

Consider the network system in the Figure 7.9 below. All components have equal failure probability $p = 5 \cdot 10^{-2}$. The system fails when there is no connection between the source and terminal nodes.

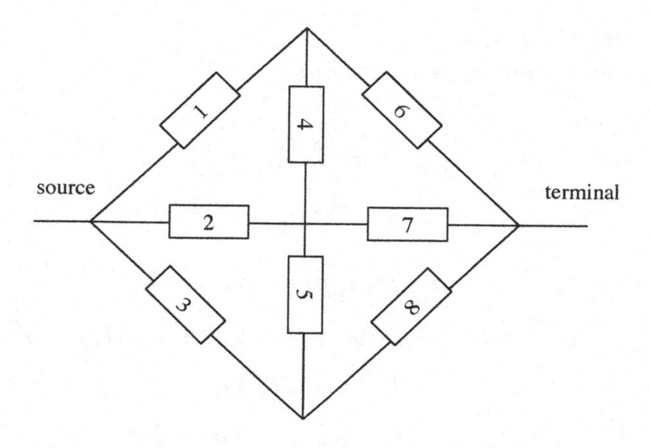


Figure 7.9. Network system

1. Identify the minimal cut sets of the network system.
2. Evaluate system unreliability from the minimal cut sets found in 1.

**Solution**

1. Minimal cut sets

We find the minimal cut sets directly by inspection from the system structure. The network system fails due to the interruption of the connection between the source and the terminal nodes.

$$M_1 = \{1,2,3\} \qquad M_5 = \{2,3,4,6\}$$
$$M_2 = \{6,7,8\} \qquad M_6 = \{3,5,6,7\}$$
$$M_3 = \{1,4,7,8\} \qquad M_7 = \{1,3,4,5,7\}$$
$$M_4 = \{1,2,5,8\} \qquad M_8 = \{2,4,5,6,8\}$$

2. System unreliability
Using the rare event approximation:

$$U_{mcs} \cong \sum_i P(M_i)$$

Where,

$$P(M_1) = P(M_2) = p^3$$
$$P(M_3) = P(M_4) = P(M_5) = P(M_6) = p^4$$
$$P(M_7) = P(M_8) = p^5$$
$$U_{mcs} \cong 2p^3 + 4p^4 + 2p^5$$
$$U_{mcs} \cong 1.25 \cdot 10^{-4}$$

**7.6 Fault tree and minimal cut sets**

Consider the following fault tree:

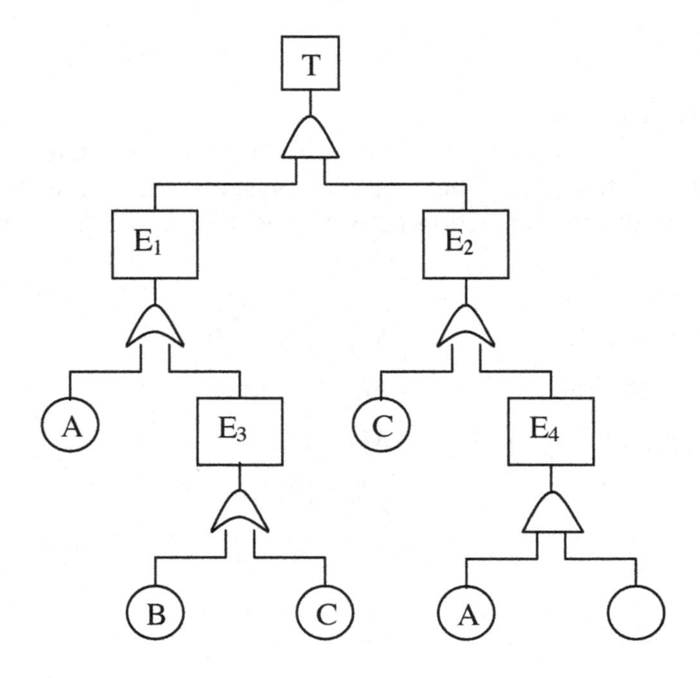

$P(A) = 0.05$, $P(B) = 0.03$, $P(C) = 0.005$.

1. Write the system structure function.
2. Reduce the structure function to find the minimal cut sets.
3. Compute the probability of the top event working through the fault tree.
4. Compute the probability of the top event solving the structure function in 1.
5. Compute the probability of the top from the minimal cut sets found in 2.

**Solution**

1. System structure function

$$X_T = X_{E_1} X_{E_2}$$
$$X_{E_1} = 1-(1-X_A)(1-X_{E_3})$$
$$X_{E_3} = 1-(1-X_B)(1-X_C) = X_B + X_C - X_B X_C$$
$$X_{E_2} = 1-(1-X_C)(1-X_{E_4})$$
$$X_{E_4} = X_A X_B$$

Substituting the above expressions in equation for $X_T$, we have:

$$X_T = (X_A + X_B + X_C - X_B X_C - X_A X_B - X_A X_C + X_A X_B X_C)(X_C + X_A X_B - X_A X_B X_C)$$

2. Minimal cut-sets
The above expression can be reduced by doing the multiplication in the right-hand term and applying the rules of Boolean logic. By doing so, the expression can be simplified as:

$$X_T = X_C + X_A X_B - X_A X_B X_C = 1-(1-X_C)(1-X_A X_B)$$

which explicitly identifies the two minimal cut sets of the system, $M_1$, $M_2$:

$$M_1 = X_A X_B$$
$$M_2 = X_C$$

3. Probability of top event working through the fault tree
Repeated components appear in the fault tree. Therefore, the computation of the failure probability $P(X_T = 1)$ working through the fault tree yields to wrong results. In fact, it has to be recalled that this approach is equivalent to calculating the failure probability by means of the non reduced expression in eq.:

$$X_T = (X_A + X_B + X_C - X_B X_C - X_A X_B - X_A X_C$$
$$+ X_A X_B X_C)(X_C + X_A X_B - X_A X_B X_C)$$

The value (wrong) thereby obtained is $5.40 \cdot 10^{-4}$.

4. Probability of top event solving the structure function
The failure probability obtained from the reduced system structure function in is:

$$P(X_T = 1) = E[X_T] = E[X_C] + E[X_A]E[X_B]$$
$$- E[X_A]E[X_B]E[X_C] = 6.4925\, 10^{-3}$$

5. Probability of top event from the minimal cut-sets found in 2
The probabilities of the basic events are very low, so that the typical rare event approximation can be used:

$$P\left(\Phi\left(X_A, X_B, X_C\right) = 1\right) \cong \sum_{j=1}^{mcs} P\left(M_j\right) = E[X_C] + E[X_A]E[X_B] = 6.5\, 10^{-3}$$

It can be seen that the approximation introduced has negligible effects.

## 7.7 Electrical generating system

An electrical generating system is shown in the figure below in block diagram form. Only the major components are to be considered: the engines $E_1$, $E_2$, and the generators $G_1$, $G_2$, $G_3$. Each generator is rated at 30 KVA. The system is required to supply at least 60KVA.

1. Draw a fault tree for the failure of the system to satisfy the required demand.
2. Find the minimal cut sets.
3. Estimate the unreliability of the system for one month operation given that the failure rate for each engine is $5\ 10^{-6}\ h^{-1}$ and for each generator $10^{-5}\ h^{-1}$.

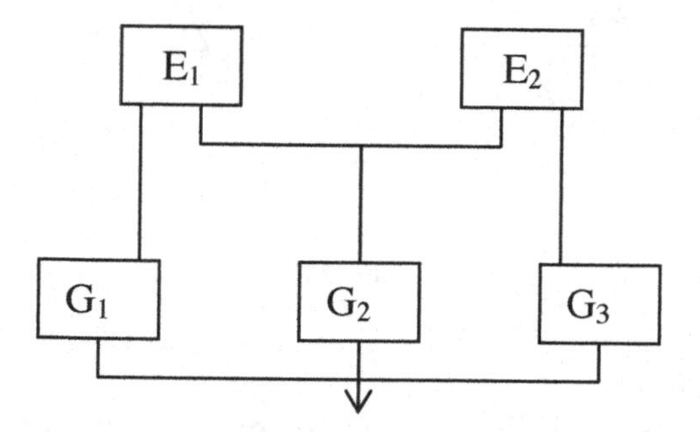


**Solution**

1. System fault tree

The fault tree is built on the bases of the following considerations. The systems fails to provide the required power (60KVA) if at least two out of the three generators do not work. In this case the supplied power is 30KVA or lower. Then, the causes for the failure of each generator to provide the power are analyzed. The failures for generators $G_1$ or $G_3$ require the primary failure of the components or of the corresponding feeding engine $E_1$ or $E_2$, respectively (branches A and C in the tree). The failure of the generator $G_2$ occurs upon primary failure of the generator and upon failure of both engines $E_1$ or $E_2$ (branch B in the tree).

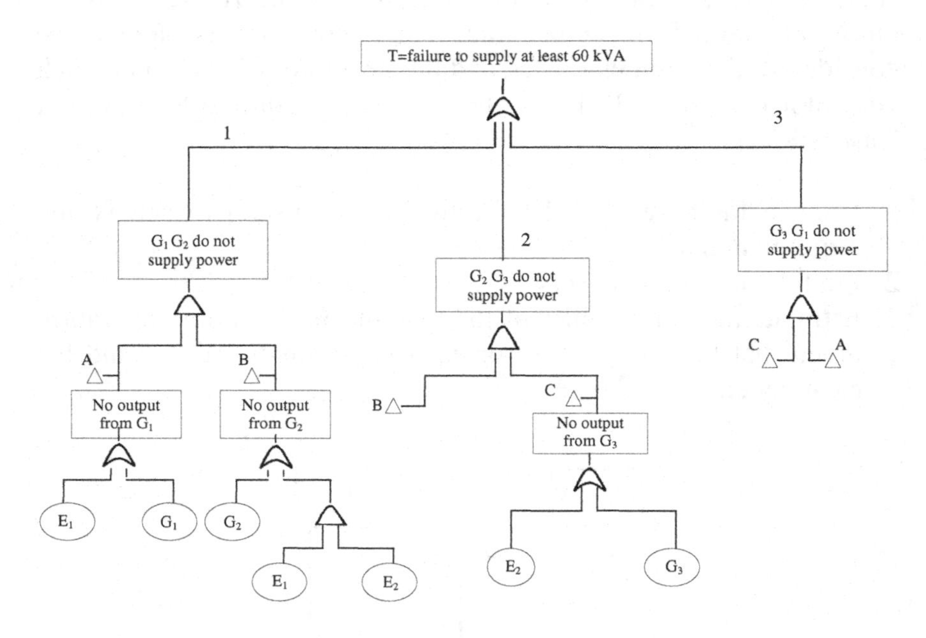


Figure 7.10.

2. Minimal cut-sets

For simplicity, we will deduce the system cut sets directly from the fault tree. We consider the three branches 1, 2, 3 of the tree one at a time.

**Subtree 1***:*

We can identify 4 cut sets, not necessarily minimal:

$$M_1 = \{E_1, G_2\}$$

$$M_2 = \{E_1, E_2\}$$

$M_3 = \{G_1, G_2\}$

$M_4 = \{G_1, E_1, E_2\}$

**Subtree 2**:

With an analogous procedure, we get for the subtree 2:

$M_5 = \{G_2, E_2\}$

$M_6 = \{G_2, G_3\}$

$M_7 = \{E_1, E_2\}$

$M_8 = \{E_1, E_2, G_3\}$

**Subtree 3**:

$M_9 = \{E_1, E_2\}$

$M_{10} = \{G_1, E_2\}$

$M_{11} = \{G_1, G_3\}$

$M_{12} = \{E_1, G_3\}$

If only the minimal cut sets are considered:

$M_1 = \{E_1, G_2\}$ $M_5 = \{G_2, E_2\}$

$M_2 = \{E_1, E_2\}$ $M_6 = \{E_1, G_3\}$

$M_3 = \{G_1, G_2\}$ $M_7 = \{G_1, E_2\}$

$M_4 = \{G_2, G_3\}$ $M_8 = \{G_1, G_3\}$

3. System reliability

The components have exponentially distributed failure rates so that the probabilities of failure within 1 month = 720 h equal:

$p_E = 1 - e^{-5 \cdot 10^{-6} \cdot 720} = 3.6E-3$, for the two engines

$p_G = 1 - e^{-10^{-5} \cdot 720} = 7.2E-3$, for the three generators.

By resorting to a first order, rare event approximation for the system unreliability:

$$P(\Phi = 1) \cong \sum_{j=1}^{mcs} P(M_j) = 2.7 \cdot 10^{-4}.$$

Where:

$$P(M_1) = 3.6 \cdot 10^{-3} \cdot 7.2 \cdot 10^{-3} = 2.59 \cdot 10^{-5} = P(M_5) = P(M_6) = P(M_7)$$
$$P(M_2) = 3.6 \cdot 10^{-3} \cdot 3.6 \cdot 10^{-3} = 1.3 \cdot 10^{-5}$$
$$P(M_3) = 7.2 \cdot 10^{-3} \cdot 7.2 \cdot 10^{-3} = 5.2 \cdot 10^{-5} = P(M_4) = P(M_8)$$

**7.8 Emergency cooling system**

The following system is designed to deliver emergency cooling to a nuclear reactor.

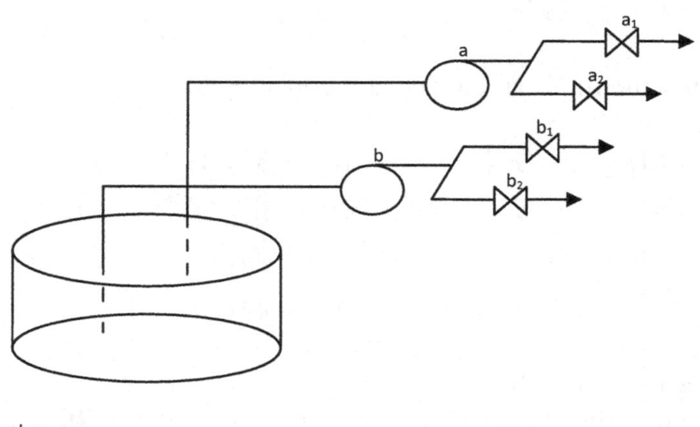

In the event of an accident the protection system delivers an actuation signal to the two identical pumps and the four identical valves. The pumps then start up, the valves open, and liquid coolant is delivered to the reactor. The following failure probabilities are found to be significant:

$p_{ps} = 10^{-5}$ the probability that the protection system will not deliver a signal to the pump valve actuators.
$p_p = 2 \cdot 10^{-2}$ the probability that a pump will fail to start when the actuation signal is received.
$p_v = 10^{-1}$ the probability that a valve will fail to open when the actuation signal is received.
$p_r = 0.5 \cdot 10^{-5}$ the probability that the reservoir will be empty at the time of the accident.

1. Draw a fault tree for the failure of the system to deliver any coolant to the primary system in the event of an accident.
2. Write the system structure function.
3. Reduce the structure function to find the minimal cut sets.
4. Compute the probability of the top event working through the fault tree.
5. Compute the probability of the top event solving the structure function in 2.
6. Compute the probability of the top event from the minimal cut sets found in 3.

**Solution**

1. Fault tree

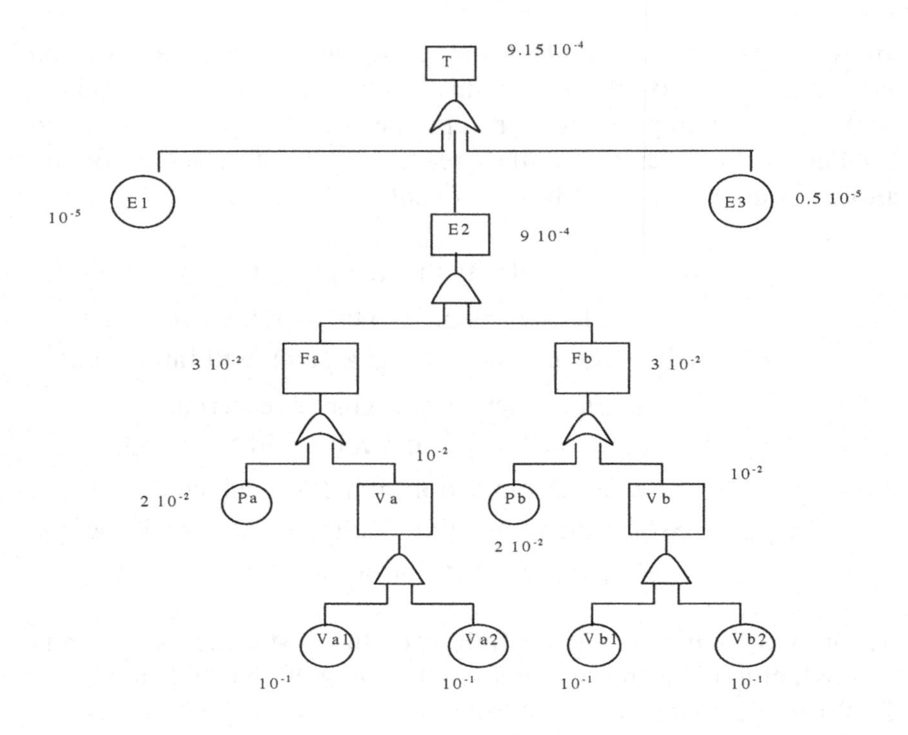


Figure 7.11.

T: no coolant deliver to the primary system in case of an accident
E1: no signal delivered to the pumping system
E2: pumping system fails to start
E3: reservoir empty
Fa: pumping system A fails to start
Fb: pumping system B fails to start
Pa: pump a fails to start
Pb: pump b fails to start
Va: valves a fail to start
Vb: valves b fail to start
Va1: valve $a_1$ fails to start
Va2: valve $a_2$ fails to start
Vb1: valve $b_1$ fails to start
Vb2: valve $b_2$ fails to start

2. System structure function

The system structure function $\Phi(\underline{X}) \equiv X_T$, which expresses the state of the system indicator variable $X_T$ as a function of the vector of the indicator variables of the basic failure events $\underline{X} \equiv (X_{E1}, X_{E3}, X_{Pa}, X_{Pb}, X_{Va1}, X_{Va2}, X_{Vb1}, X_{Vb2})$, is obtained as follows:

$$\Phi(\underline{X}) = X_T = 1-(1-X_{E1})(1-X_{E2})(1-X_{E3})$$
$$X_{E2} = X_{Fa}X_{Fb}$$
$$X_{Fa} = 1-(1-X_{Pa})(1-X_{Va}) = 1-(1-X_{Pa})(1-X_{Va1}X_{Va2})$$
$$X_{Fb} = 1-(1-X_{Pb})(1-X_{Vb}) = 1-(1-X_{Pb})(1-X_{Vb1}X_{Vb2})$$

Substituting the above expressions in the structure function yields:

$$\Phi(\underline{X}) = X_T = 1-(1-X_{E1})(1-X_{E3})\begin{Bmatrix} 1-[1-(1-X_{Pa})(1-X_{Va1}X_{Va2})] \\ \cdot[1-(1-X_{Pb})(1-X_{Vb1}X_{Vb2})] \end{Bmatrix}$$

3. Minimal cut-sets

We firstly focus on the term $X_{Fa}X_{Fb}$. The step for the successive groupings of the terms down to the minimal cut sets are:

$$\begin{aligned} X_{Fa}X_{Fb} &= [1-(1-X_{Pa})(1-X_{Va1}X_{Va2})]\cdot[1-(1-X_{Pb})(1-X_{Vb1}X_{Vb2})] \\ &= [1-(1-X_{Pa})(1-X_{Va})]\cdot[1-(1-X_{Pb})(1-X_{Vb})] \\ &= (X_{Pa}+X_{Va}-X_{Pa}X_{Va})(X_{Pb}+X_{Vb}-X_{Pb}X_{Vb}) \\ &= X_{Pa}X_{Pb}+X_{Pa}X_{Vb}-X_{Pa}X_{Pb}X_{Vb}+X_{Va}X_{Pb} \\ &\quad +X_{Va}X_{Vb}-X_{Va}X_{Pb}X_{Vb}-X_{Pa}X_{Va}X_{Pb} \\ &\quad -X_{Pa}X_{Va}X_{Vb}+X_{Pa}X_{Va}X_{Pb}X_{Vb} \\ &= 1-1+X_{Pa}X_{Pb}+X_{Pa}X_{Vb}-X_{Pa}X_{Pb}X_{Vb}+X_{Va}X_{Pb} \\ &\quad +X_{Va}X_{Vb}-X_{Va}X_{Pb}X_{Vb}-X_{Pa}X_{Va}X_{Pb}-X_{Pa}X_{Va}X_{Vb} \\ &\quad +X_{Pa}X_{Va}X_{Pb}X_{Vb}+X_{Pa}X_{Va}X_{Pb}X_{Vb}-X_{Pa}X_{Va}X_{Pb}X_{Vb} \end{aligned}$$

$$
\begin{aligned}
&= 1-[(1-X_{Pa}X_{Vb}-X_{Va}X_{Pb}-X_{Va}X_{Vb}+X_{Va}X_{Pb}X_{Vb}\\
&\quad X_{Pa}X_{Va}X_{Pb})-X_{Pa}X_{Pb}(1-X_{Pa}X_{Vb}-X_{Va}X_{Pb}-X_{Va}X_{Vb}\\
&\quad +X_{Va}X_{Pb}X_{Vb}+X_{Pa}X_{Va}X_{Pb})]\\
&= 1-(1-X_{Pa}X_{Pb})(1-X_{Pa}X_{Vb}-X_{Va}X_{Pb}-X_{Va}X_{Vb}\\
&\quad +X_{Va}X_{Pb}X_{Vb}+X_{Pa}X_{Va}X_{Pb})\\
&= 1-(1-X_{Pa}X_{Pb})(1-X_{Pa}X_{Vb}-X_{Va}X_{Pb}-X_{Va}X_{Vb}\\
&\quad +X_{Va}X_{Pb}X_{Vb}+X_{Pa}X_{Va}X_{Pb}+X_{Pa}X_{Va}X_{Pb}X_{Vb}\\
&\quad -X_{Pa}X_{Va}X_{Pb}X_{Vb})\\
&= 1-(1-X_{Pa}X_{Pb})[(1-X_{Va}X_{Pb}-X_{Va}X_{Vb}+X_{Va}X_{Pb}X_{Vb})\\
&\quad -X_{Pa}X_{Vb}(1-X_{Va}X_{Pb}-X_{Va}X_{Vb}+X_{Va}X_{Pb}X_{Vb})]\\
&= 1-(1-X_{Pa}X_{Pb})(1-X_{Pa}X_{Vb})(1-X_{Va}X_{Pb}-X_{Va}X_{Vb}\\
&\quad +X_{Va}X_{Pb}X_{Vb})\\
&= 1-(1-X_{Pa}X_{Pb})(1-X_{Pa}X_{Vb})(1-X_{Va}X_{Vb})(1-X_{Va}X_{Pb})
\end{aligned}
$$

$$
\begin{aligned}
X_{Fa}X_{Fb} &= 1-(1-X_{Pa}X_{Pb})(1-X_{Pa}X_{Vb1}X_{Vb2})\\
&\cdot(1-X_{Va1}X_{Va2}X_{Vb1}X_{Vb2})(1-X_{Va1}X_{Va2}X_{Pb})
\end{aligned}
$$

Thus:

$$
\begin{aligned}
\Phi(\underline{X}) = X_T &= 1-(1-X_{E1})(1-X_{E2})(1-X_{Pa}X_{Pb})\\
&\cdot(1-X_{Pa}X_{Vb1}X_{Vb2})(1-X_{Va1}X_{Va2}X_{Vb1}X_{Vb2})\\
&\cdot(1-X_{Va1}X_{Va2}X_{Pb})
\end{aligned}
$$

By so doing, $N_{mcs}=6$ minimal cut sets are identified:

$M_1 = \{E1\}$ $\quad$ $M_4 = \{Pa, Vb1, Vb2\}$

$M_2 = \{E3\}$ $\quad$ $M_4 = \{Va1, Va2, Vb1, Vb2\}$

$M_3 = \{Pa, Pb\}$ $\quad$ $M_6 = \{Pb, Va1, Va2\}$

4. Probability of top event working through the fault tree
Multiplying the probabilities of the events through the fault tree, as indicated in the Figure, one readily obtains:

$$P(X_T = 1) = 9.15 \cdot 10^{-4}$$

Repeated components do not appear in the different branches of the fault tree. Therefore, the computation of the failure probability $P(X_T = 1)$ working through the fault tree yields a correct result.

5. Probability of the top event solving the structure function in 2.
Applying the expectation operator to the structure function in 2,

$$P(X_T = 1) = 1 - (1 - E[X_{E1}])(1 - E[X_{E3}])\{1 - \{1 - (1 - E[X_{Pa}])$$
$$(1 - E[X_{Va1}] \cdot E[X_{Va2}])\}\{1 - (1 - E[X_{Pb}])(1 - E[X_{Vb1}] \cdot E[X_{Vb2}]\}\}$$

$$P(X_T = 1) = 9.03 \cdot 10^{-4}$$

Note that we can use directly the equation in 2 because repeated components do not appear in the fault tree. Otherwise, the correct probability would have to be obtained from the reduced form of the structure function equation in 3.

6. Probability of the top event solving the structure function in 3.
By resorting to a first order, rare event approximation:

$$P(X_T = 1) \cong \sum_{j=1}^{N_{mcs}} P(M_j) = P(E1) + P(E2) + P(Pa)P(Pb)$$
$$+P(Pa)P(Vb1)P(Vb2) + P(Va1)P(Va2)P(Vb1)P(Vb2)$$
$$+P(Pb)P(Va1)P(Va2)$$

Then,

$$P(X_T = 1) = 9.15 \cdot 10^{-4}$$

**7.9 Network system 2**

Consider the network system shown in Figure 7.12. All the components have equal failure rate $\lambda = 10^{-4}$ days$^{-1}$. The system fails when there is no connection between node I and node O. We also consider the nodes as perfect, i.e. they cannot fail.

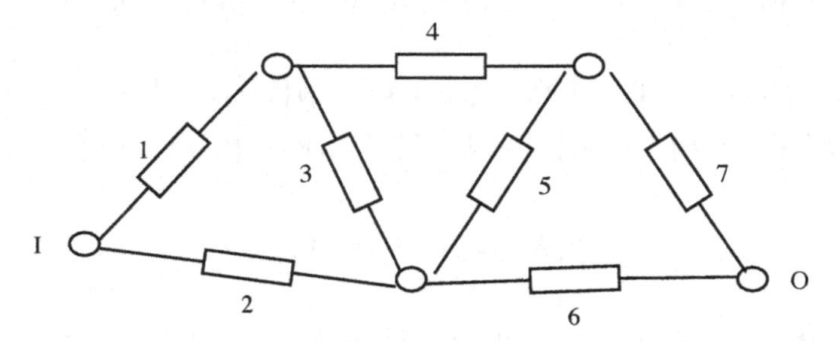


Figure 7.12.

1. For the network of the figure above, develop a fault tree for the event "no signal at O given a signal at I". Neglect human errors and external causes of secondary failures.
2. Identify the minimal cut-sets of the network system.
3. Evaluate analytically the system unreliability at the mission time $T_M$ of 1 year from the minimal cut-sets found above.
4. Evaluate the system unreliability at the mission time $T_M$ of 10 year from the minimal cut-sets found in 3.

## Solution

### 1. Fault tree

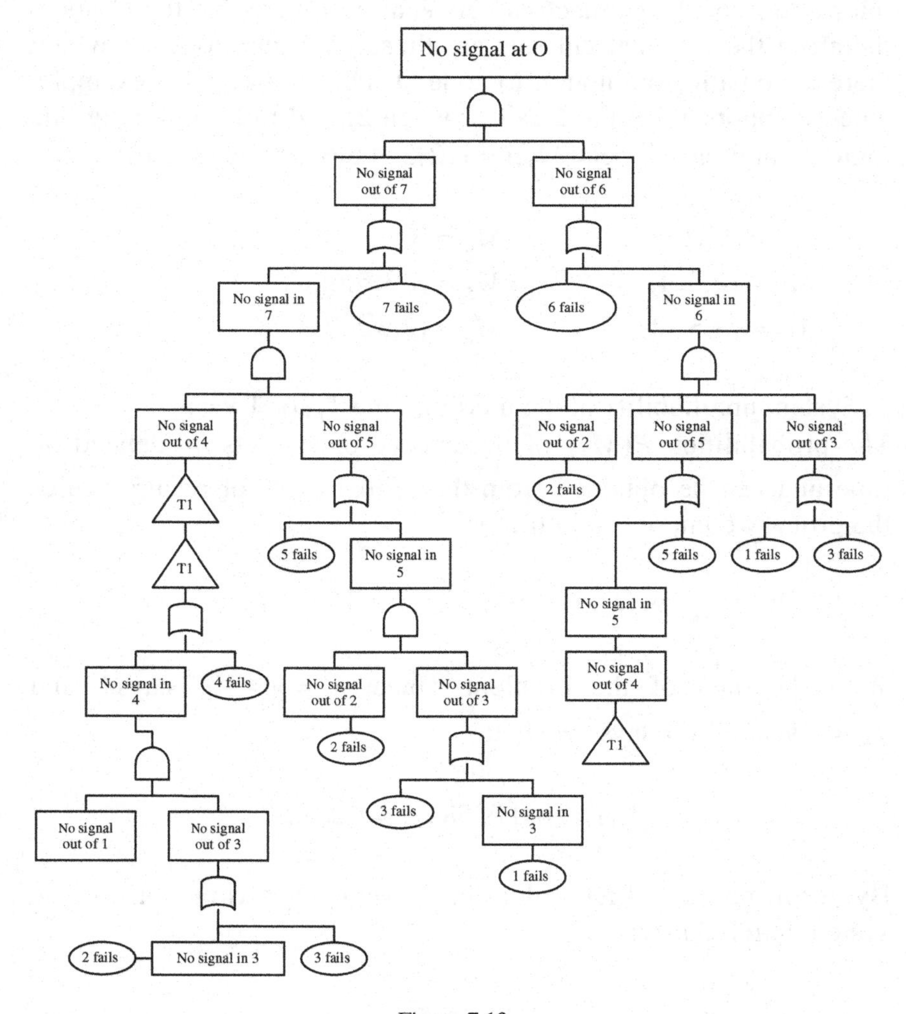


Figure 7.13.

2. Minimal cut-sets

Instead of using the fault tree, we find the minimal cut sets directly from the system structure. The network system fails due to the interruption of the connection between the I node and the O node, therefore the minimal cut sets are those configurations for which there is no path from node I to node O. This happens, for example, when components 1 and 2 fail or when 2, 3, 4 fail. Following this logic, 6 minimal cut sets $M_i$, $i = 1,2,3,4,5,6$ can be identified:

$$M_1 = \{1,2\} \qquad M_4 = \{6,7\}$$
$$M_2 = \{2,3,4\} \qquad M_5 = \{1,3,5,6\}$$
$$M_3 = \{4,5,6\} \qquad M_6 = \{2,3,5,7\}$$

3. System unreliability at the mission time $T_M$ of 1 year

The probabilities $P(M_i)$ of occurrence of cut sets $M_i$ depend on time and can be obtained from the probabilities of occurrence of the primary components' failures

$$p_k(T_M) = 1 - e^{-\lambda_k T_M}, \quad k = 1,2,...,7 .$$

With the data of the problem, since $\lambda_k = \lambda = 10^{-4}$ days$^{-1}$ and $T_M = 1 \text{ year} = 365 \text{ days}$, we get:

$$p_k(T_M) = 0.0358, \quad k = 1,2,...,7$$

By resorting to a first order, rare event approximation for the system unreliability:

$$U_{mcs}(T_M) = \sum_{j=1}^{N} P(M_j) = 2.7 \cdot 10^{-4}$$

4. System unreliability at the mission time $T_M$ of 10 year
With the new value of $T_M = 10\text{ years} = 10 \cdot 365\text{ days}$ we have a higher value of the component's failures probabilities:

$$p_k(T_M) = 0.305\,,\; k = 1,2,...,7$$

By resorting to a first order, rare event approximation for the system unreliability:

$$U_{mcs}(T_M) = \sum_{j=1}^{N} P(M_j) = 0.2617$$

If a higher precision in the $U_{mcs}(T_M)$ value is required, we can compute a second order approximation which gives a lower limit for the unreliability:

$$U_{mcs}(T_M) = \sum_{i=1}^{N} P(M_i) - \sum_{i=2}^{N}\sum_{j=1}^{i-1} P\{M_i M_j\} = 0.2090$$

Thus: $$\sum_{i=1}^{N} P(M_i) - \sum_{i=2}^{N}\sum_{j=1}^{i-1} P\{M_i M_j\} < U_{mcs}(T_M) < \sum_{j=1}^{N} P(M_j)$$

### 7.10 Unavailability

A system is successful, if at least two of four nominally identical elements are "up". The probability that one element is "down" is denoted by $q$.

1. Find the minimal cut sets (mcs).
2. Using these mcs derive an exact expression for the system unavailability, i.e., the probability that the system is "down".

3. Find the system unavailability using the binomial distribution and compare with the result of part 2.
4. How would parts 1., 2., and 3. be affected (i.e. would they still be applicable?), if the elements were not nominally identical (i.e. different probabilities $q_i$, $i=1,2,3,4$)? Simply give arguments without carrying out any calculations.

**Solution**

1. Minimal cut-sets

$$M_1 = \{X_1, X_2, X_3\} \qquad M_3 = \{X_2, X_3, X_4\}$$
$$M_2 = \{X_1, X_2, X_4\} \qquad M_4 = \{X_1, X_3, X_4\}$$

2. System unavailability using mcs

$$\begin{aligned} T &= 1-(1-X_1X_2X_3)(1-X_1X_2X_4)(1-X_2X_3X_4)(1-X_1X_3X_4) \\ &= X_1X_2X_3 + X_1X_2X_4 + X_2X_3X_4 + X_1X_3X_4 - 6\cdot X_1X_2X_3X_4 \\ &\quad +4\cdot X_1X_2X_3X_4 - X_1X_2X_3X_4 \\ &= X_1X_2X_3 + X_1X_2X_4 + X_2X_3X_4 + X_1X_3X_4 - 3\cdot X_1X_2X_3X_4 \end{aligned}$$

The system unavailability therefore is $Q = 4q^3 - 3q^4$.

3. System unavailability using binomial distribution

$$\begin{aligned} Q &= P(\text{system is down}) = P(\text{exactly 3 elements are down}) \\ &\quad + P(\text{4 elements are down}) \\ &= B(3 \mid 4, q) + B(4 \mid 4, q) \end{aligned}$$

$$= \binom{4}{3} q^3 (1-q) + q^4$$
$$= 4q^3(1-q) + q^4$$
$$= 4q^3 - 3q^4$$

Same as in 2.

4. Elements not nominally identical
For non-identical elements, the *mcs* approach still works, however the binomial approach cannot be used.

**7.11 Sensored alarm system**

Three sensors are to be used to provide a reliable warning of whether a system is on or off. Each sensor output is connected to an indicator light. When it is working properly, each sensor activates the indicator light as soon as the system starts up, and extinguishes the light at the moment the system is shut down. The indicator lights are considered perfect, but with probability $r$ it is possible for the operator to misinterpret the control panel of the three sensor lights. Suppose that each sensor may fail in either of two modes:

(1) it may fail "on", so that its output lights the indicator lamp regardless of whether the system is on or off, the probability of this failure being $q_{on}$;

(2) it may fail "off", so that the indicator lamp is out whether the system is on or off, the probability of this failure being $q_{off} = 1 - p - q_{on}$, where $p$ is the probability of sensor success.

There are two operators:

Operator A judges the system to be on if he or she perceives one, two, or three indicator lights are on and that it is off if all lights are out.
Operator B judges the system to be on if he or she perceives two or three indicator lights are on and the system to be off if two or three of the indicator lights are out.

From the viewpoint of A and B:

1. Write the fault trees for the event: *indicator lights on when system off*.
2. Write the fault trees for the event: *indicator lights off when system on*.
3. Write the system structure functions for 1 and 2.
4. Reduce the structure functions to find the minimal cut sets for 1 and 2.
5. Compute the probability of the top event working through the fault trees of 1 and 2.
6. Compute the probability of the top event solving the structure functions in 3.
7. Compute the probability of the top event from the minimal cut sets found in 4.
8. If $p = 0.88$, $q_{on} = 0.02$, $q_{off} = 0.10$ and $r = 0.10$, discuss the system performance from the point of view of Safety and of Reliability/Availability: which operator is best from the two different points of view?

## Solution

# 1. "indicator lights on when system off"

$A_1$: Operator A

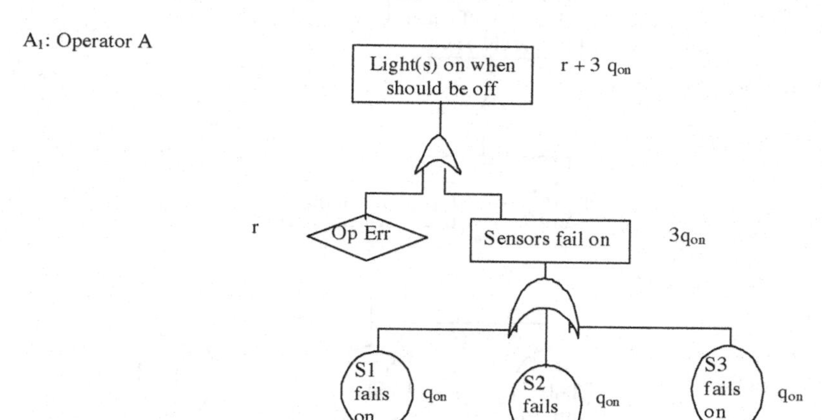


$B_1$: Operator B

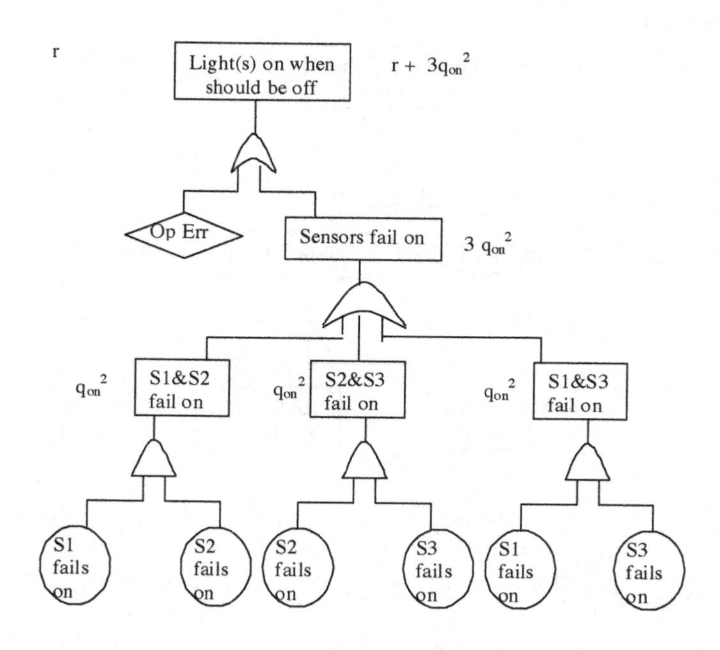

# 2. "indicator lights off when system on"

$A_2$: Operator A

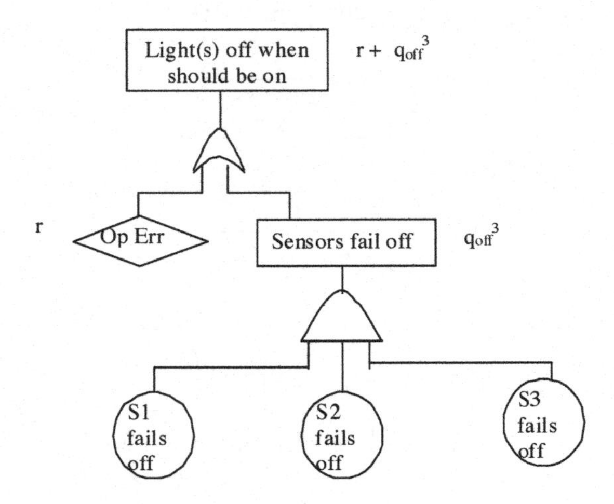


$B_2$: Operator B

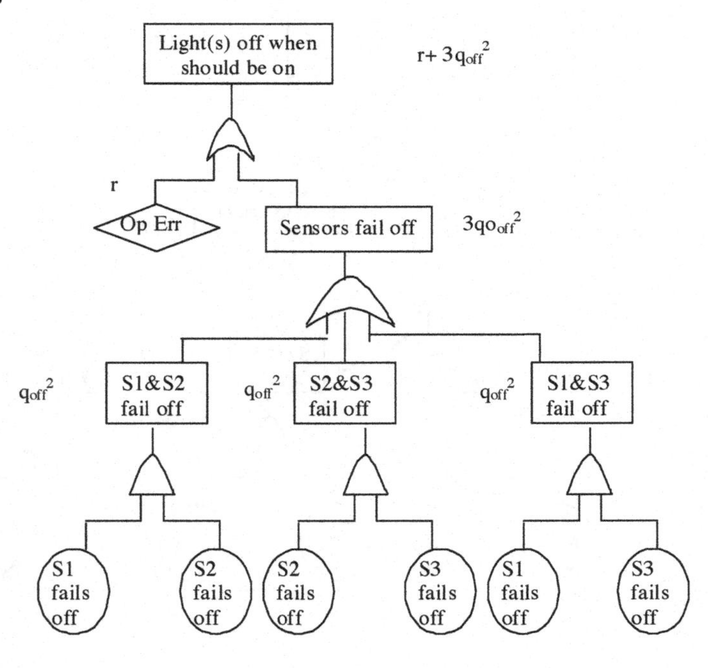

3. System structure functions for 1 and 2

$X_E$ = indicator variable for the event "*operator error*"
$X_{ion}$ = indicator variable for the event "*i-th sensor fails on*", $i = 1,2,3$
$X_{ioff}$ = indicator variable for the event "*i-th sensor fails off*", $i = 1,2,3$

A1:

$$\Phi(\underline{X}) = X_T = 1-(1-X_E)\{1-[1-(1-X_{1on})(1-X_{2on})(1-X_{3on})]\}$$
$$= 1-(1-X_E)(1-X_{1on})(1-X_{2on})(1-X_{3on})$$

B1:

$$\Phi(\underline{X}) = X_T = 1-(1-X_E)\{1-[1-(1-X_{1on}X_{2on})$$
$$\cdot(1-X_{2on}X_{3on})(1-X_{1on}X_{3on})]\}$$
$$= 1-(1-X_E)(1-X_{1on}X_{2on})(1-X_{2on}X_{3on})(1-X_{1on}X_{3on})$$

A2:

$$\Phi(\underline{X}) = X_T = 1-(1-X_E)(1-X_{1off}X_{2off}X_{3off})$$

B2:

$$\Phi(\underline{X}) = X_T = 1-(1-X_E)\{1-[(1-(1-X_{1off}X_{2off})$$
$$\cdot(1-X_{2off}X_{3off})(1-X_{1off}X_{3off})]\}$$
$$= 1-(1-X_E)(1-(1-X_{1off}X_{2off})(1-X_{2off}X_{3off})(1-X_{1off}X_{3off})$$

4. Minimal cut-sets

A1:

$$M_1 = \{X_E\},\ M_2 = \{X_{1on}\},\ M_3 = \{X_{2on}\},\ M_4 = \{X_{3on}\}$$

B1:

$$M_1 = \{X_E\},\ M_2 = \{X_{1on}, X_{2on}\},\ M_3 = \{X_{2on}, X_{3on}\},\ M_4 = \{X_{1on}, X_{3on}\}$$

A2:

$$M_1 = \{X_E\},\ M_2 = \{X_{1off}, X_{2off}, X_{3off}\}$$

B2:

$$M_1 = \{X_E\},\ M_2 = \{X_{1off}, X_{2off}\},\ M_3 = \{X_{2off}, X_{3off}\},\ M_4 = \{X_{1off}, X_{3off}\}$$

5. Probability of the top event working through the fault tree
Considering the rare event approximation and multiplying the probabilities of the events through the fault tree, as indicated in the figures, one readily obtains:

**A1**: $P(X_T = 1) \cong r + 3q_{on}$ **B1**: $P(X_T = 1) \cong r + 3q_{on}^2$
**A2**: $P(X_T = 1) \cong r + q_{off}^3$ **B2**: $P(X_T = 1) \cong r + 3q_{off}^2$

6. Probability of the top event solving the structure function in 3

**A1**: $P(X_T = 1) = 1 - (1-r)(1-q_{on})(1-q_{on})(1-q_{on})$
**B1**: $P(X_T = 1) = 1 - (1-r)(1-q_{on}q_{on})(1-q_{on}q_{on})(1-q_{on}q_{on})$

Repeated components appear in the fault tree for B1. Therefore the computation of the probability $P(X_T = 1)$ through the structure functions of for B1 in 3. yields to a wrong result. To obtain the correct failure probability we should first expand all the products of the structure function of B1 and reduce the system structure function.

**A2**: $P(X_T = 1) = 1 - (1-r)(1-q_{off}q_{off}q_{off})$

**B2**: $P(X_T = 1) = 1 - (1-r)(1 - q_{off} q_{off})(1 - q_{off} q_{off})(1 - q_{off} q_{off})$

Repeated components appear in the fault tree for B2. Therefore the computation of the probability $P(X_T = 1)$ through the structure function of B2 yields to a wrong result. To obtain the correct failure probability we should first expand all the products of the equation of B2 and reduce the system structure function.

7. Probability of the top event from the minimal cut-sets found in 4
Using the rare event approximation:

**A1**:

$$P(X_T = 1) \approx P(X_E = 1) + P(X_{1on} = 1) + P(X_{2on} = 1)$$
$$+P(X_{3on} = 1) \approx r + 3q_{on} \approx 0.16$$

**B1**:

$$P(X_T = 1) \approx P(X_E = 1) + P(X_{1on} = 1)P(X_{2on} = 1)$$
$$+P(X_{2on} = 1)P(X_{3on} = 1)$$
$$+P(X_{1on} = 1)P(X_{3on} = 1) \approx r + 3q_{on}^2 \approx 0.101$$

**A2**:

$$P(X_T = 1) \approx P(X_E = 1) + P(X_{1off} = 1)P(X_{2off} = 1)$$
$$P(X_{3off} = 1) \approx r + q_{off}^3 \approx 0.101$$

**B2**:

$$P(X_T = 1) \approx P(X_E = 1) + P(X_{1off} = 1)P(X_{2off} = 1)$$
$$+P(X_{2off} = 1)P(X_{3off} = 1)$$
$$+P(X_{1off} = 1)P(X_{3off} = 1) \approx r + 3q_{off}^2 \approx 0.13$$

8. Discuss which operator is the best
**Safety**: It is important to know when the system is operating, to avoid dangerous situations.

$$P_{A2}(\text{system believed off}|\text{system actually on}) <$$
$$P_{B2}(\text{system believed off}|\text{system actually on})$$

A is better

**Availability**: It is important to reduce false alarms (one thinks that the system is producing when, actually, it is not).

$$P_{A1}(\text{system believed on}|\text{system actually off}) >$$
$$P_{B1}(\text{system believed on}|\text{system actually off})$$

B is better

### 7.12 Domestic hot water system

In the domestic hot water system in the Figure 7.16, the control of the temperature is achieved by the controller opening and closing the main gas valve when the water temperature goes outside the preset limits $T_{\min} = 140\,\text{F}$, $T_{\max} = 180\,\text{F}$.

1. Formulate a list of undesired safety and reliability events.
2. Construct the fault tree for the top event *rupture of water tank* assuming only the following primary failure events.
3. For this event, write the system structure function.
4. Reduce the structure function to find the minimal cut sets.
5. Assume primary failure event probabilities equal to 0.1 and compute the probability of the top event working through the fault tree.
6. Compute the probability of the top event solving the structure function in 6.
7. Compute the probability of the top event from the minimal cut sets found in 7.

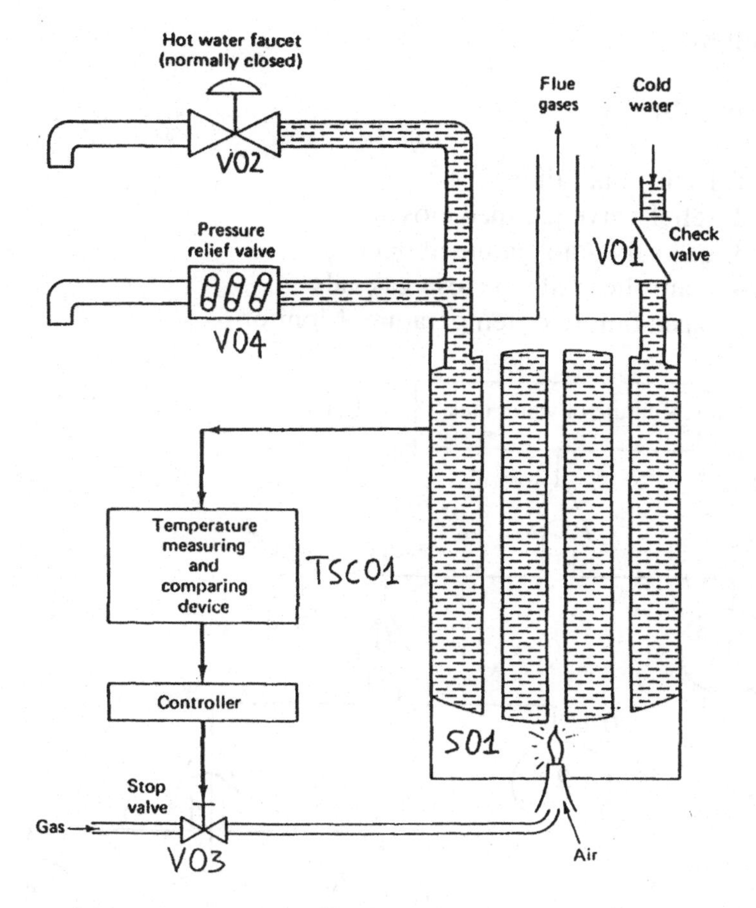


Figure 7.14.

## Solution

1. Undesired safety and reliability events
   - Tank rupture (safety)
   - Water too cold (reliability)
   - Water too hot (safety/reliability)
   - Insufficient water flow (reliability)
   - Excessive flow (reliability)

2. Fault tree

Basic events:

1: basic tank failure
2: relief valve jammed closed
3: gas valve fails jammed open
4: controller fails to close gas valve
5: basic failure of temperature. Monitor

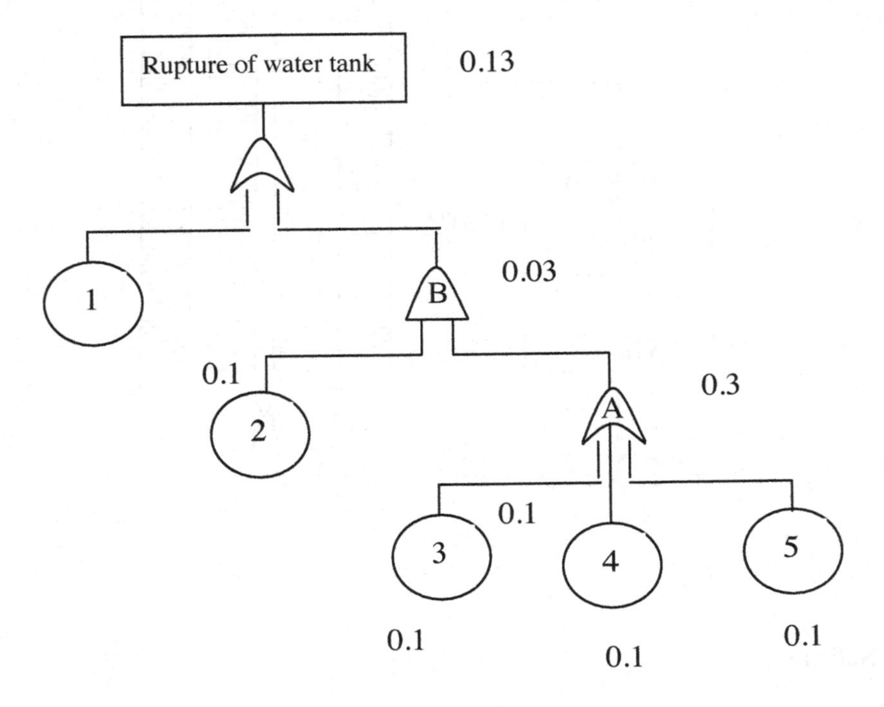


3. Structure function
The structure function $X_T \equiv \Phi(X_1, X_2, X_3, X_4, X_5)$ is obtained as follows:

$$X_T = 1-(1-X_1)(1-X_B)$$
$$X_T = 1-(1-X_1)(1-X_2X_A)$$
$$X_T = 1-(1-X_1)\{1-X_2[1-(1-X_3)(1-X_4)(1-X_5)]\}$$
$$X_T = 1-(1-X_1)\{1-X_2[1-(1-X_3)(1-X_4-X_5+X_4X_5)]\}$$
$$X_T = 1-(1-X_1)\{1-X_2[X_3+X_4+X_5-X_4X_5$$
$$-X_3X_5-X_3X_4+X_3X_4X_5)]\}$$
$$X_T = 1-(1-X_1)(1-X_2X_3-X_2X_4-X_2X_5+X_2X_3X_4$$
$$+X_2X_3X_5+X_2X_4X_5-X_2X_3X_4X_5)$$

By successive manipulation of the last term of the above equation, we can reduce the structure function to find the minimal cut sets.

$$X_T = 1-(1-X_1)[(1-X_2X_4-X_2X_5+X_2X_4X_5)-X_2X_3(1-X_2X_4$$
$$-X_2X_5+X_2X_4X_5)]$$
$$X_T = 1-(1-X_1)(1-X_2X_3)(1-X_2X_4-X_2X_5+X_2X_4X_5)$$
$$X_T = 1-(1-X_1)(1-X_2X_3)[(1-X_2X_5)-X_2X_4(1-X_2X_5)]$$
$$X_T = 1-(1-X_1)(1-X_2X_3)(1-X_2X_5)(1-X_2X_4)$$

4. Minimal cut-sets

$$M_1 = \{X_1\} \qquad M_3 = \{X_2, X_4\}$$
$$M_2 = \{X_2, X_3\} \qquad M_4 = \{X_2, X_5\}$$

5. Probability of the top event working through the fault tree
Multiplying the probabilities of the events through the fault tree, as indicated in the Figure, one readily obtains:

$$P(X_T = 1) = 0.13$$

6. Probability of the top event solving the structure function
Applying the expectation operator to the structure function in 3. we obtain:

$$P(X_T = 1) = 0.127$$

7. Probability of the top event from the minimal cut-sets
From the minimal cut sets and the approximation of rare events, we have:

$$P(X_T = 1) = \sum_i P(M_i) = 0.1 + 0.1 \cdot 0.1 + 0.1 \cdot 0.1 + 0.1 \cdot 0.1 = 0.13$$

## Chapter 8

# Event tree analysis

### 8.1 Coolant system

Consider the coolant system in Figure 8.1.

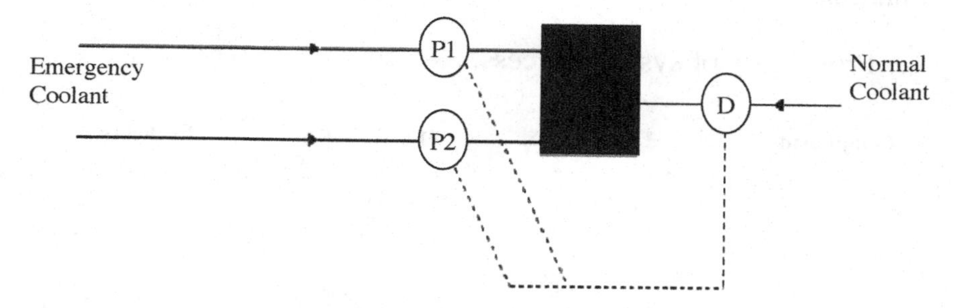


Figure 8.1. Coolant system

- P1 and P2 are electrically driven pumps, D is a flow detector, and EP (not shown) is the electric power
- Initiating event is a break in the normal coolant pipe
- Full system success (S) requires both pumps, the detection system, and the electrical power operating
- One pump operating results in partial success (P)
- Two pumps failing or failure of electrical power (EP) results in system failure (F)

Numerical values are given in Table 8.1.

Table 8.1. Numerical values

| Component | R: reliability | Q: probability of failure |
|---|---|---|
| P1 | $R_{P1} = 0.95$ | $Q_{P1} = 0.05$ |
| P2 | $R_{P2} = 0.95$ | $Q_{P2} = 0.05$ |
| D | $R_D = 0.96$ | $Q_D = 0.04$ |
| EP | $R_{EP} = 0.98$ | $Q_{EP} = 0.02$ |

Find the probability of system success.

**Solution**

1. Probability of system success.

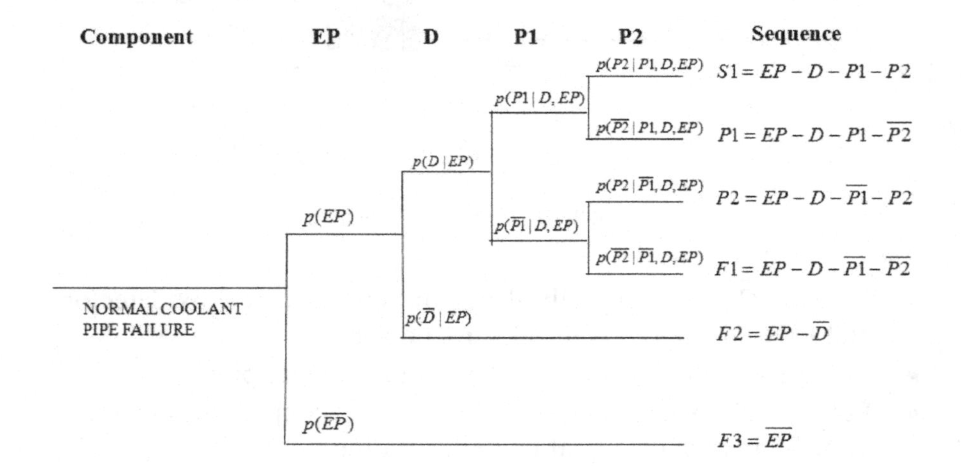


Figure 8.2. Event tree

$$p(S1) = p(EP)p(D \mid EP)p(P1 \mid D, EP)p(P2 \mid P1, D, EP)$$
$$= R_{EP}R_D R_{P1} R_{P2} = 0.849$$
$$p(P1) = p(EP)p(D \mid EP)p(P1 \mid D, EP)p(\overline{P2} \mid P1, D, EP)$$
$$= R_{EP}R_D R_{P1} Q_{P2} = 0.045$$

$$p(P2) = p(EP)\,p(D \mid EP)\,p(\overline{P1} \mid D,EP)\,p(P2 \mid P1,D,EP)$$
$$= R_{EP}R_{D}Q_{P1}P_{P2} = 0.045$$

The probability of system success is given by:

$$R_s = p(S1) + p(P1) + p(P2) = 0.939$$

**8.2 Electric pump system**

An electric pump is fed by a group of three electric generators. Given the high power requested by the pump, it is necessary that at least two of the generators be in operation.

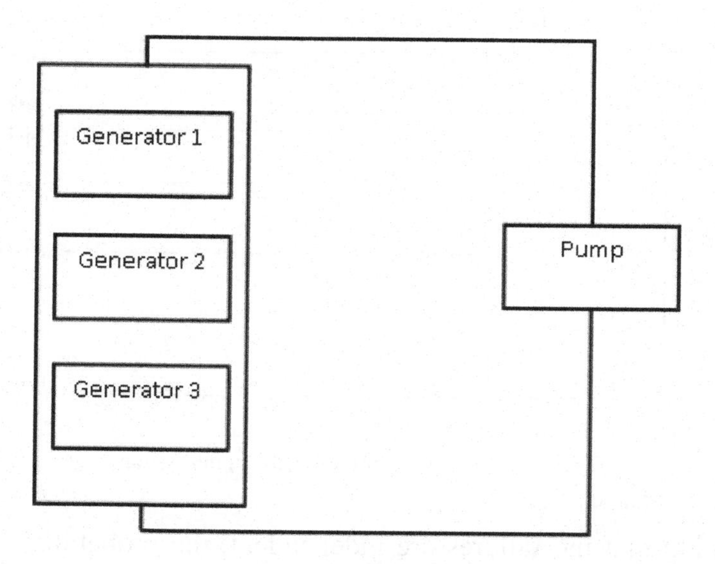


Figure 8.3. Electric pump system

Build the event tree to calculate the probability of operation of the system knowing the data in Table 8.2.

Table 8.2. Data of the electric pump system

| Component | Reliability | Prob. of failure |
|---|---|---|
| Pump | 0.90 | 0.10 |
| Generator 1 (G1) | 0.85 | 0.15 |
| Generator 2 (G2) | 0.85 | 0.15 |
| Generator 3 (G3) | 0.85 | 0.15 |

## Solution

1. Probability of the system operating.

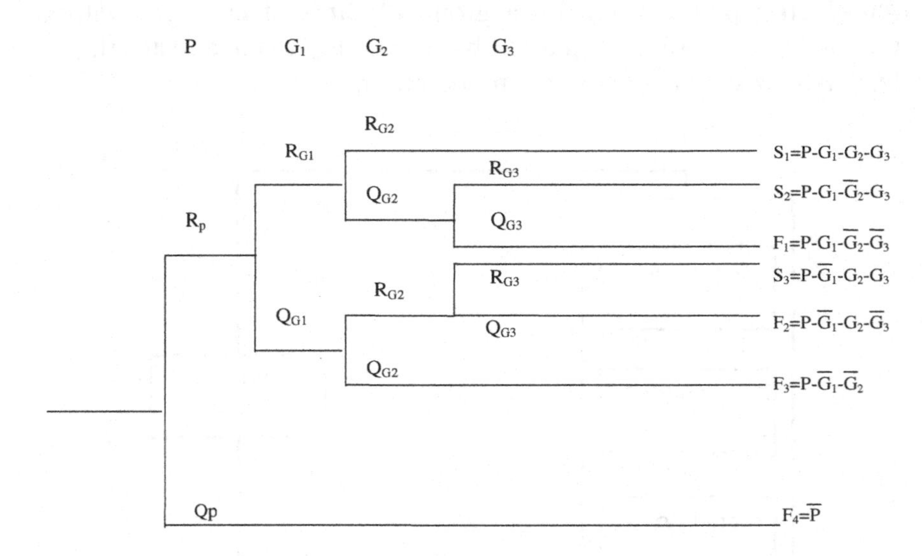


Figure 8.4. Event tree for the electric pump system

Assuming that the failures are independent, the probability of the system operating is given by:

$$R_S = p(S1) + p(S2) + p(S3)$$
$$= (R_p R_{G1} R_{G2}) + (R_p R_{G1} Q_{G2} R_{G3}) + (R_p Q_{G1} R_{G2} R_{G3}) = 0.845$$

## 8.3 Lamp supply system

The system represented in Figure 8.5 illustrates the operation of a lamp fed by two batteries and a power unit. In order to have energy in the circuit it is enough that one among the two batteries and the power unit works.

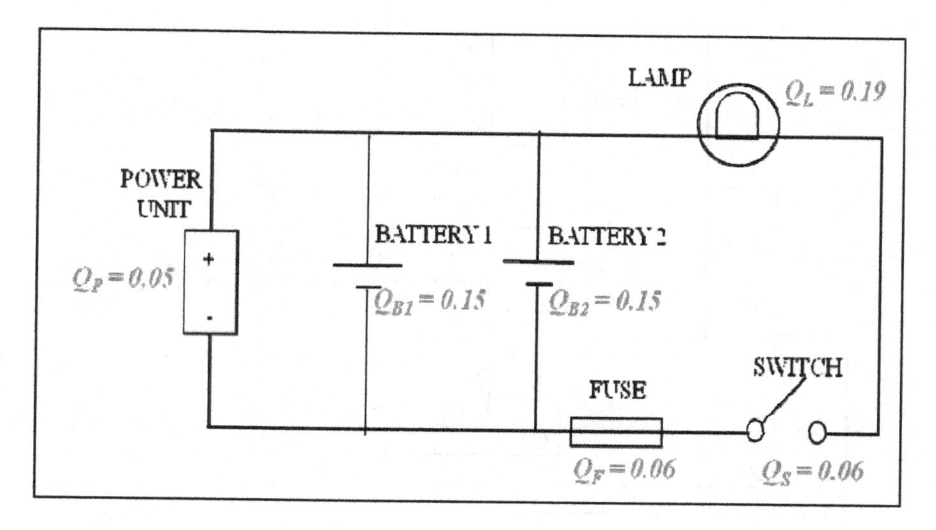


Figure 8.5. Electrical circuit of the lighting system

Build the event tree for the event "failure of the lighting system" and compute its probability based on the component failure probabilities indicated in Figure 8.5.

### Solution

1. Probability of the event "failure of the lighting system".

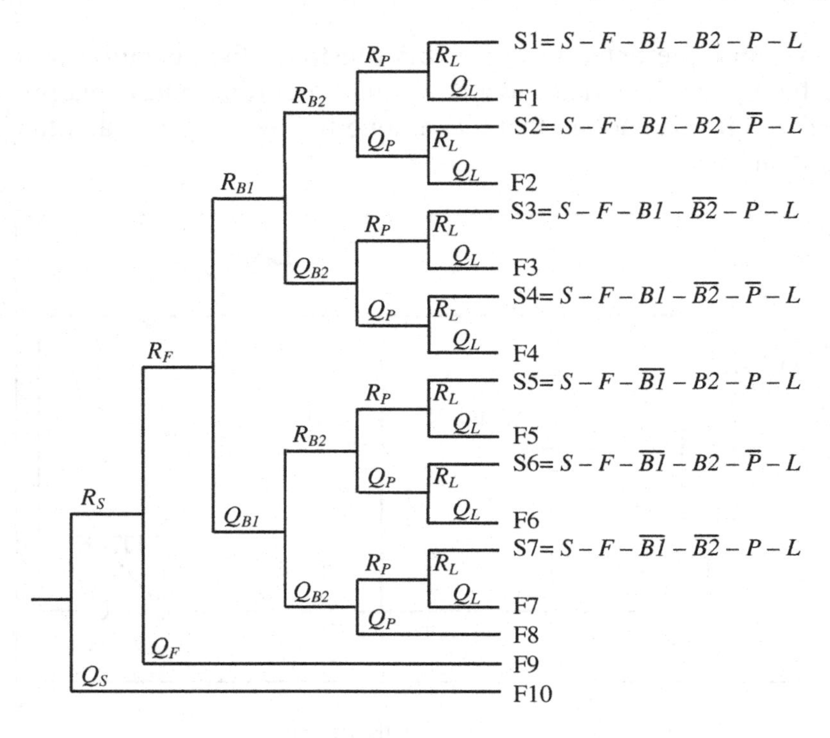


Figure 8.6. Event tree

$$R_{SYS} = p(S1) + p(S2) + p(S3) + p(S4) + p(S5) + p(S6) + p(S7)$$
$$= 0.715$$

$$Q_{SYS} = 1 - R_{SYS} = 0.285$$

## 8.4 Poisonous gas deposit

Let us consider a person working in a deposit of poisonous gas. The deposit is subject to leakages which are controlled by a gas detector. If a leakage occurs, an alarm sounds and the person should leave the place immediately although not all gas leakages necessarily imply the presence of gas in the work place. Failure can be due to the detector not detecting the gas, the alarm not sounding or the person not leaving the facility. The main concern is if the worker is wounded or not.

Assuming that the failure sequences are mutually exclusive, find the probability of the worker being wounded.

**Solution**

Probability of the worker being wounded.

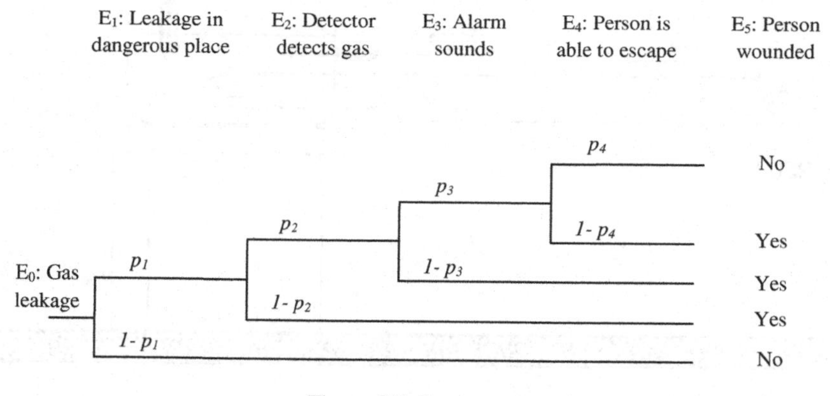


Figure 8.7. Event tree

$$p(\text{worker being wounded}) = p_1(p_2(p_3(1-p_4)+(1-p_3))+(1-p_2))$$

## 8.5 Electric fryer system

Let us consider the electric fryer system showed in Figure 8.8 with the following legend:

1. Electric fryer
2. Oil
3. Thermostat
4. High temperature switch
5. Smoke detector
6. Sprinkler

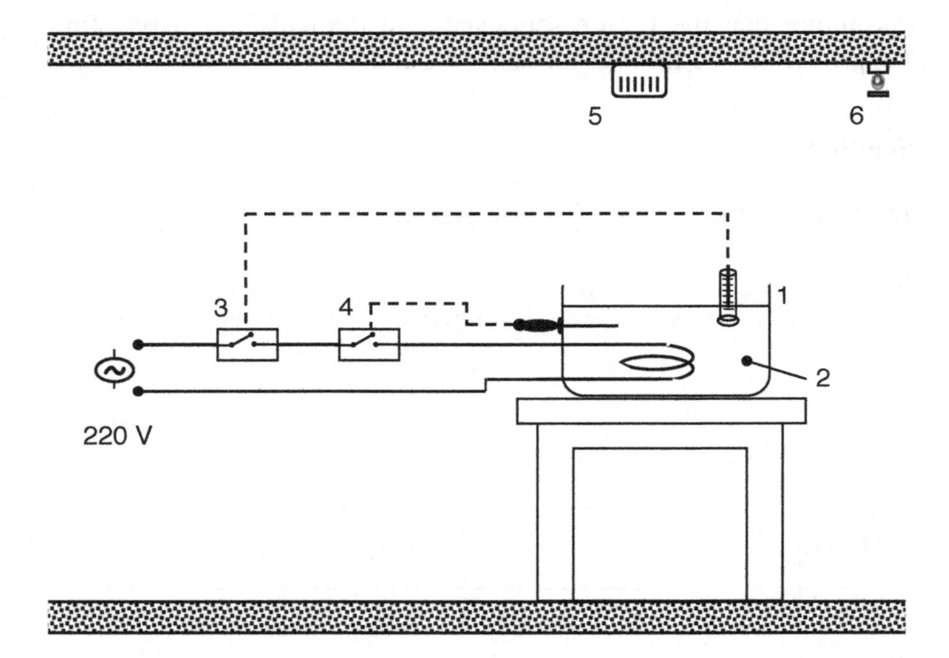


Figure 8.8. Electric fryer

Construct the event tree starting from the event "thermostat jammed closed".

**Solution**

1. Event tree

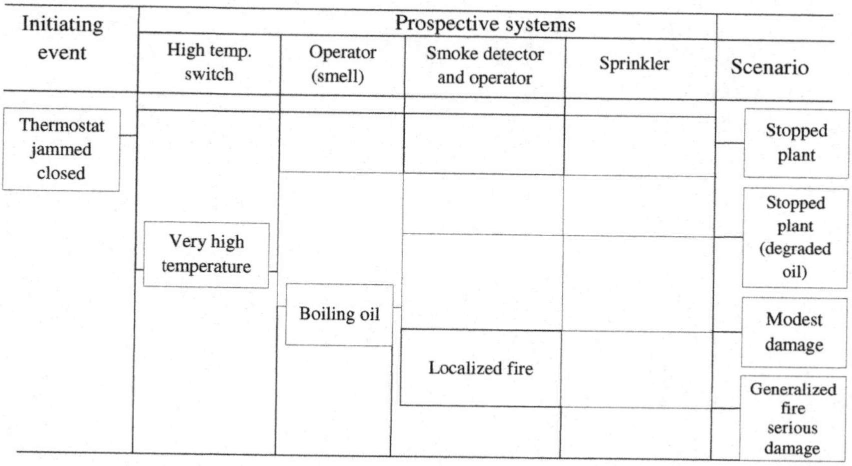

| Initiating event | Prospective systems | | | | Scenario |
|---|---|---|---|---|---|
| | High temp. switch | Operator (smell) | Smoke detector and operator | Sprinkler | |
| Thermostat jammed closed | | | | | Stopped plant |
| | Very high temperature | | | | Stopped plant (degraded oil) |
| | | Boiling oil | | | Modest damage |
| | | | Localized fire | | Generalized fire serious damage |

Figure 8.9. Event tree

# Chapter 9

# Estimation of reliability parameters from experimental data

## 9.1 Failure times

The failure time data (5.2, 6.8, 11.2, 16.8, 17.8, 19.6, 23.4, 25.4, 32.0, 44.8 minutes) are exponentially distributed as

$$F_X(x) = 1 - e^{-\frac{x}{\vartheta}}:$$

Make a probability plot and estimate the parameter, $\vartheta$.

### Solution

Probability plot and parameter estimate

We have: $\ln\left[\frac{1}{1-F(x)}\right] = \frac{x}{\theta}$.

We then approximate $F_X(x_i)$ by $\hat{F}(x_i) = \frac{i}{N+1}$ where $i$ = 1, 2, 3, …, $N$.

In our case, since $N = 10$, we have

$$\ln\left[\frac{1}{1-F(t_i)}\right] = \frac{11}{(11-i)}$$

$$= [1.1, 1.222, 1.373, 1.571, 1.833, 2.2, 2.75, 3.666, 5.5, 11]$$

In the next Figure 9.1, these numbers have been plotted on semilog paper versus the failure times. After drawing a straight line through the data we note that when $\dfrac{1}{1-F} = e = 2.72$, then $\hat{\theta} = x = 21\,\text{min}$.

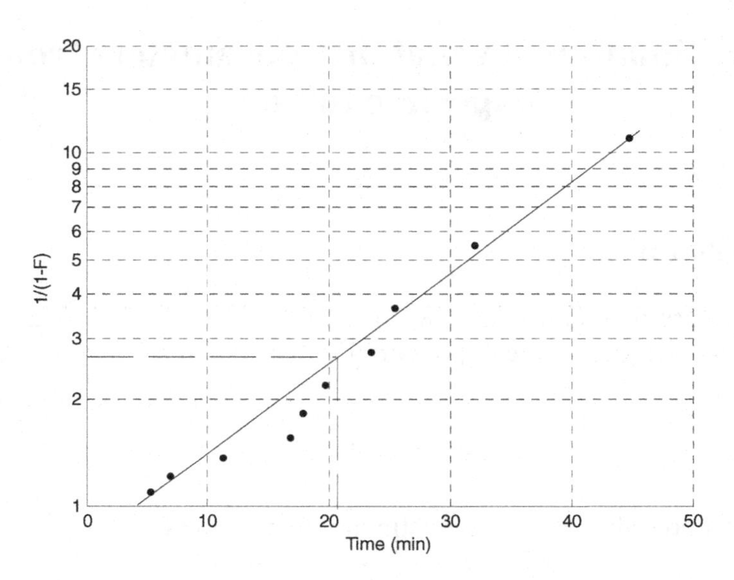


Figure 9.1. Probability plot of exponential distributed data

## 9.2 Catalytic converter test

Twenty units of a catalytic converter are tested to failure without censoring. The times to failure (in days) are the following

Table 9.1. Times of failures

| | | | | |
|---|---|---|---|---|
| 2.6 | 3.2 | 3.4 | 3.9 | 5.6 |
| 7.1 | 8.4 | 8.8 | 8.9 | 9.5 |
| 9.8 | 11.3 | 11.8 | 11.9 | 12.7 |
| 12.3 | 16.0 | 21.9 | 22.4 | 24.2 |

1. Plot on exponential paper and determine whether the failure rate is increasing or decreasing with time.

2. Plot the results on Weibull paper and estimate its parameters.
3. Find the method-of-moments estimates of the Weibull parameters.

**Solution**

1. Exponential paper

$$F(t) = 1 - e^{\left(-\frac{t}{\tau}\right)^{\beta}} = 1 - e^{-\lambda(t)t} \text{ then } \lambda(t) = \frac{1}{\tau}\left(\frac{1}{\tau}\right)^{\beta-1}$$

If $\beta > 1$ then $\lambda(t)$ increasing. If $\beta < 1$ then $\lambda(t)$ decreasing. From Figure 9, $\lambda(t)$ is the slope and seems to be slightly increasing, therefore, we expect $\beta > 1$.

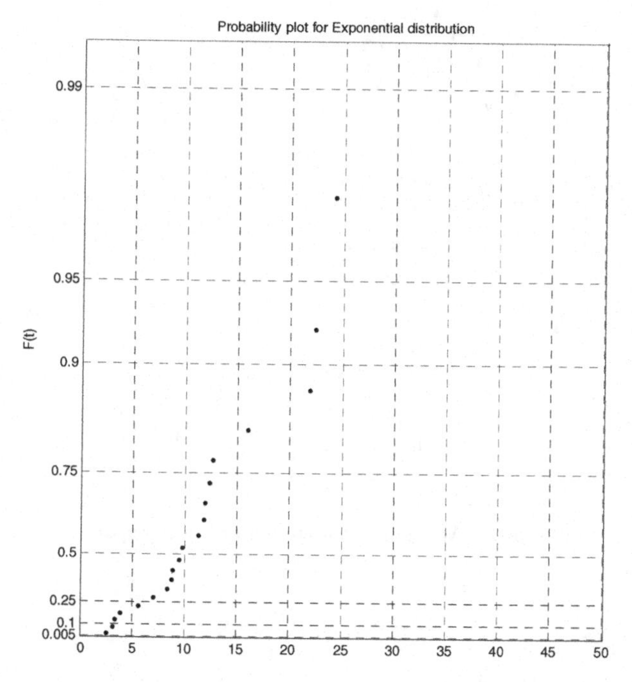


Figure 9.2. Exponential distribution probability paper

## 2. Weibull Paper

The Weibull cumulative distribution function is: $F(t) = 1 - e^{\left(-\frac{t}{\tau}\right)^{\beta}}$. Setting $t = \tau$, we get $F(\tau) = 0.632$. Looking at Figure 9.3, we get $\tau = 12$. The slope $\beta \approx 1.1$.

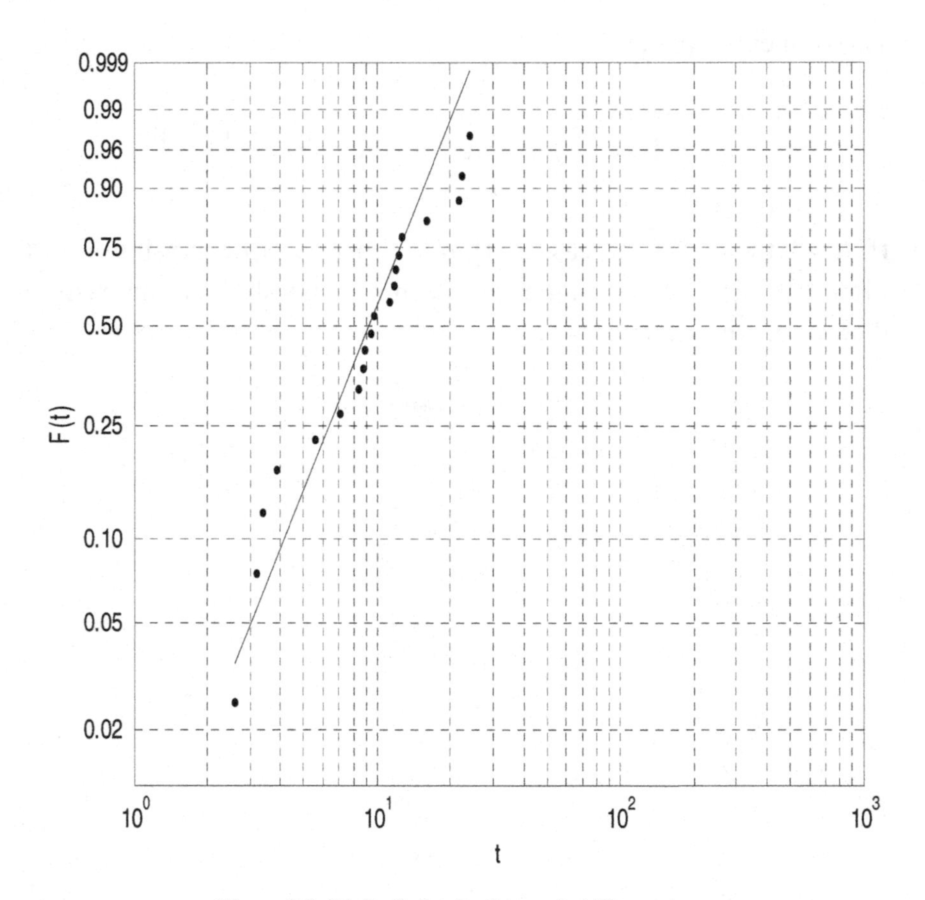


Figure 9.3. Weibull distribution probability paper

3. Method of Moments

$$\overline{X} = \frac{\sum_{i=1}^{20} X_i}{20} = 10.785 = \tau\left[\Gamma\left(\frac{1}{\beta}+1\right)\right] \Rightarrow \tau = \frac{10.785}{\Gamma\left(1+\frac{1}{\beta}\right)} = \frac{10.785\beta}{\Gamma\left(\frac{1}{\beta}\right)}$$

$$s^2 = \frac{1}{19}\left[\sum_{i=1}^{20} x_i^{\,2} - 20\times 10.785^2\right] = 39.834$$

$$= \tau^2\left[\Gamma\left(\frac{2}{\beta}+1\right) - \Gamma\left(\frac{1}{\beta}+1\right)\right]^2$$

$$\Rightarrow \tau^2\left[\frac{2}{\beta}\Gamma\left(\frac{2}{\beta}\right) - \frac{1}{\beta}\Gamma\left(\frac{1}{\beta}\right)\right]^2 = 39.834$$

The problem reduces to finding two unknowns from two equations. Solving for $\tau$ and $\beta$ we get:

$$\tau \approx 11.51 \text{ and } \beta \approx 1.2$$

### 9.3 Confidence bounds

Suppose that the time to failure $T$ (years) of a certain item is an exponential random variable with probability density function:

$$h(t) = \lambda e^{-\lambda t},\ t > 0$$

1. If we have a sample of 3 observations on $T$, i.e. $E = \{t_1 = 1, t_2 = 2.8, t_3 = 2.2\}$, find the 95% upper confidence bound and the 90% confidence interval for $\Lambda$ using frequentist statistics.

2. Find the 95% upper confidence bound and the 90% confidence interval for $\Lambda$ using frequentist statistics and using the information in 1. What are the corresponding Bayesian quantities?

**Solution**

1. The 95% upper confidence bound and the 90% confidence interval.

The evidence $E = \{t_1 = 1, t_2 = 2.8, t_3 = 2.2\}$ constitutes the result of an uncensored test, but we can consider the test as a type 2 censoring test (test ends at the $r$-th item failure) with a total time to failure $T = 1 + 2.2 + 2.8 = 6$ and $r = 3$.
Using the Table for the normal standard variate we compute:

The 95% upper bound:

$$\left(\frac{2T}{z_{95,2r}}\right)^{-1} = \left(\frac{2*6}{z_{95,6}}\right)^{-1} = \left(\frac{12}{12.6}\right)^{-1} = 1.05\,years^{-1}$$

where $Z_{100\varepsilon,\nu}$ denotes the upper $\varepsilon$ percentile of the chi-square distribution with $\nu$ degrees of freedom.

The 90% confidence interval for $\Lambda$:

$$P\left(\left(\frac{2T}{z_{05,6}}\right)^{-1} \le \lambda \le \left(\frac{2T}{z_{95,6}}\right)^{-1}\right) = 0.9$$

$$P\left(\left(\frac{2*6}{z_{05,6}}\right)^{-1} \le \lambda \le \left(\frac{2*6}{z_{95,6})}\right)^{-1}\right) = 0.9$$

$$P(0.1367 < \lambda \le 1.05) = 0.9$$

2. Censored test of 4 components
Now the evidence is regarded as the result of a Type I Censoring test (test ends at a fixed $t_0$). We have tested $m = 4$ items, with $n = 3$ failures before $t_0 = 2.8s$. Thus the total time to test $T$ is:

$$T = \sum_i t_i + (m - n)t_0 = (1 + 2.2 + 2.8) + (4 - 3) \cdot 2.8 = 8.8s$$

The 95% upper bound is:

$$\left(\frac{2T}{z_{95,2r+2}}\right)^{-1} = \left(\frac{2 \cdot 8.8}{z_{95,8}}\right)^{-1} = \left(\frac{17.6}{15.5}\right)^{-1} = 0.8807\, years^{-1}$$

where $Z_{100\varepsilon,\nu}$ denotes the upper $\varepsilon$ percentile of the chi-square distribution with $\nu$ degrees of freedom.

The 90% confidence interval for $\Lambda$:

$$P\left(\left(\frac{2T}{z_{05,2r}}\right)^{-1} \leq \lambda \leq \left(\frac{2T}{z_{95,2r+2}}\right)^{-1}\right) = 0.9$$

$$P\left(\left(\frac{2 * 8.8}{z_{05,6}}\right)^{-1} \leq \lambda \leq \left(\frac{2 * 8.8}{z_{95,8}\,)}\right)^{-1}\right) = 0.9$$

$$P(0.093 < \lambda \leq 1.05) = 0.9$$

*Bayesian statistics*

The 95% upper bound:

The 95% upper confidence bound for $\Lambda$ is the value $\lambda_{95}$ such that:

$$P(\Lambda \leq \lambda_{95}) = \int_{-\infty}^{\lambda_{95}} \pi''(\lambda|E)d\lambda = \int_{-\infty}^{\lambda_{95}} \frac{\nu''[\nu''\lambda]^{k''-1}}{\Gamma(k'')} e^{-\nu''\lambda} d\lambda = 0.95$$

With $k'' = 4$ and $\nu'' = 10.8$

$$\frac{10.8^4}{6} \int_{-\infty}^{\lambda_{95}} \lambda^3 e^{-10.8\lambda} d\lambda = 0.95 \Rightarrow \lambda_{95} = 0.717$$

The 90% confidence interval for $\Lambda$:

Analogously the 90% confidence interval for $\Lambda$ is an interval $[\lambda_{05}, \lambda_{95}]$ such that

$$P[\lambda_{05} < \Lambda < \lambda_{95} | E] = \int_{\lambda_{05}}^{\lambda_{95}} \pi''(\lambda|E) d\lambda = 0.90$$

We now compute $\lambda_{05}$:

$$P(\Lambda < \lambda_{05}) = \frac{10.8^4}{6} \int_{0}^{\lambda_{05}} \lambda^3 e^{-10.8\lambda} d\lambda = 0.05 \Rightarrow \lambda_{05} = 0.126$$

Thus, the 90% confidence interval for $\Lambda$ is [0.126, 0.717].
Note that the bayesian confidence interval is smaller than the frequentist confidence interval.

## 9.4 Remission times

Suppose that the remission time, in weeks, of leukaemia patients that have undergone a certain type of chemotherapy treatment is an exponential random variable having an unknown mean $\vartheta$. A group of twenty such patients are being monitored and, at present, their remission times are (in weeks)

1.2, 1.8*, 2.2, 4.1, 5.6, 8.4, 11.8*, 13.4*, 16.2, 21.7, 29*, 41, 42*, 42.4*, 49.3, 60.5, 61*, 94, 98, 99.2*,
where an * next to the data means that the patient's remission is still continuing, whereas an unstarred data point means that the remission ended at that time.
1. What is the maximum likelihood estimate of $\vartheta$.
2. What is the 90% two-sided confidence interval of $\vartheta$.

**Solution**

1. Maximum likelihood estimate
Remission time: period after the therapy when the patient does not present signs of leukemia.
*T*: remission time, random variable exponentially distributed with unknown mean $\vartheta$.
Adding the times of monitoring control, in weeks, we have $T_{control} = 702.8 weeks$.
From the total number of 20 monitored patients, 12 of them (without *) have terminated their periods of remission (they have shown signs of leukemia again).This is a type I sample, terminated at a fixed $t_0$.

For an exponential distribution, the $MTTF_{MLE} = \frac{T}{r}$, where $T$ is the total time on test and $r$ the number of "components" failed. Therefore

$$MTTF_{MLE} = \hat{\vartheta} = \frac{702.8}{12} = 58.6 weeks$$

2. Confidence interval
We have $r = 12$, then $2r = 24$

$$\alpha = 0.90; \ \frac{1-\alpha}{2} = 0.05; \ \frac{1+\alpha}{2} = 0.95$$

From the Chi-Square tables:

$$\chi^2_{0.05}(24) = 13.8$$
$$\chi^2_{0.95}(26) = 38.9$$

The confidence interval is then:

$$\frac{2T}{\chi^2_{0.95}} \leq \hat{\theta} \leq \frac{2T}{\chi^2_{0.05}} \text{ which gives } [36.13; 101.86]$$

### 9.5 Frequentist and Bayesian failure time estimation

Suppose that the time to failure $T$ (years) of a certain item is an exponential random variable with probability density function

$$h(t) = \lambda e^{-\lambda t}, \ t > 0$$

From prior experience we are led to believe that $\lambda$ is a value of an exponential random variable $\Lambda$ with probability density function

$$\pi'(\lambda) = 2e^{-2\lambda}, \ \lambda > 0$$

1. What is the marginal density of $T$ and what is the reliability of the item for a period of one year?
2. If we have a sample of n observations on $T$, i.e. $(t_1, t_2, \ldots t_n)$, show that the posterior distribution of $\Lambda$ is a gamma distribution.
3. If the sample in 2. is (1; 2.8; 2.2), find the (posterior) marginal distribution of $T$ and the reliability for one year.
4. What are the prior and posterior mean values and variances of $\Lambda$? (use the information in 3. What was the effect of the evidence?

5. Find the 95% upper confidence bound and the 90% confidence interval for $\Lambda$ using frequentist statistics (using the information in 3.). What are the corresponding Bayesian quantities?
6. Repeat 3., 4. and 5. assuming that the sample is as given in 2, but it now represents a censored test of 4 components.

**Solution**

1. Marginal density of T and reliability
Let be $g_{T,\Lambda}(t,\lambda)$ the joint probability density function of the random variables $(T,\Lambda)$, i.e. $g_{T,\Lambda}(t,\lambda)dtd\lambda$ is the joint probability of $T$ having a value between t and $t+dt$ and $\Lambda$ between $\lambda$ and $\lambda+d\lambda$. Then,

$$h(t)dt = h_T(t/\lambda)dt = \frac{g_{T,\Lambda}(t,\lambda)dtd\lambda}{\pi'(\lambda)d\lambda}$$

$$g_{T,\Lambda}(t,\lambda) = h(t/\lambda)\pi'(\lambda) = \lambda e^{-\lambda t}\cdot 2e^{-2\lambda}$$

Note that $\iint g_{T,\Lambda}(t,\lambda)dtd\lambda = 1$.

The marginal density function of $T$, $f_T(t)$ is defined as:

$$f_T(t) = \int g_{T,\Lambda}(t,\lambda)d\lambda = \int_0^\infty \lambda e^{-\lambda t}2e^{-2\lambda}d\lambda = \int_0^\infty 2\lambda e^{-\lambda(t+2)}d\lambda$$

(integrating by parts)

$$= \left|\frac{-1}{t+2}e^{-\lambda(t+2)}2\lambda\right|_0^\infty + \frac{2}{t+2}\int_0^\infty e^{-\lambda(t+2)}d\lambda$$

$$= 0 + \frac{2}{t+2}\left(\frac{-1}{t+2}e^{-\lambda(t+2)}\right)_0^\infty = \frac{2}{(t+2)^2}$$

The reliability of the item for a period of one year is:

$$R(1) = 1 - F(1) = 1 - \int_0^1 f(t)dt = \int_1^\infty f(t)dt = 1 - \left(\frac{-2}{t+2}\right)_0^1 = 1 + \frac{2}{3} - 1 = \frac{2}{3}$$

2. Posterior distribution

The distribution of the random variable $T=$ time to failure of the item, given any particular value of $\lambda$ is an exponential distribution. The likelihood $L(\lambda|E)$, of the evidence $E$ of $n$ observed times to failure $(t_1, t_2, ... t_n)$ is:

$$L(E|\lambda) = \prod_{i=1}^{n} \lambda e^{-\lambda t_i} = \lambda^n e^{-\lambda \sum_{i=1}^{n} t_i}$$

The posterior density of $\lambda$ is:

$$\pi''(\lambda | E) = \frac{\pi'(\lambda)L(E|\lambda)}{\int_0^\infty \pi'(\lambda)L(E|\lambda)d\lambda} = \frac{2e^{-2\lambda}\lambda^n e^{-\lambda \sum_{i=1}^{n} t_i}}{\int_0^\infty 2e^{-2\lambda}\lambda^n e^{-\lambda \sum_{i=1}^{n} t_i} d\lambda} = \frac{2e^{-2\lambda}\lambda^n e^{-\lambda \sum_{i=1}^{n} t_i}}{\frac{2n!}{\left(\sum_{i=1}^{n} t_i + 2\right)^{n+1}}}$$

$$= \left(\sum_{i=1}^{n} t_i + 2\right)^{n+1} \frac{\lambda^n}{n!} e^{-\lambda \sum_{i=1}^{n} t_i + 2} = \frac{\left(\sum_{i=1}^{n} t_i + 2\right)\left[\left(\sum_{i=1}^{n} t_i + 2\right)\lambda\right]^n}{n!} e^{-\lambda \sum_{i=1}^{n} t_i + 2}$$

If we remember that $\Gamma(k) = (k-1)!$ for any integer $k > 0$ and we denote $\nu'' = \sum_{i=1}^{n} t_i + 2$ and $k'' = n + 1$ we get that

$$\pi''(\lambda|E) = \frac{\nu''[\nu''\lambda]^{k''-1}}{\Gamma(k'')} e^{-\nu''\lambda}$$

Note that $\pi'(\lambda)$, the prior distribution of $\Lambda$, was a Gamma distribution with parameter $\nu' = 2$ and $k' = 1$ and that also $\pi''(\lambda)$, the posterior distribution of $\Lambda$, is a Gamma distribution with parameters

$$\nu'' = \nu' + \sum_{i=1}^{n} t_i$$
$$k'' = k' + n.$$

Hence we have confirmed that the family of Gamma distributions is conjugate to the family of exponential distributions.

3. Posterior marginal density of T and posterior reliability

Using the evidence $E = \{t_1 = 1, t_2 = 2.8, t_3 = 2.2\}$, we compute the posterior marginal distribution of $T$ as:

$$f(t|E) = \int h(t|\lambda)\pi''(\lambda|E)d\lambda$$

Since Exponential and Gamma are conjugate:

$$f(t|E) = \frac{\nu^k k}{(\nu + t)^{k+1}} = \frac{[(1+2.8+2.2)+2]^{(3+1)} \cdot 4}{\{[(1+2.8+2.2)+2]+t\}^{(3+1+1)}} = \frac{8^4 \cdot 4}{(8+t)^5}$$
$$= \frac{16384}{(8+t)^5}$$

The posterior reliability of the item for a period of 1 year is:

$$R(1) = 1 - F(1) = 1 - \int_0^1 f(t)dt = \int_1^\infty \frac{16384}{(8+t)^5} dt = 16384\left(\frac{-1}{4}\frac{1}{(8+t)^4}\right)_1^\infty$$
$$= \frac{16384}{4*1^4} = 0.6243$$

4. Prior and posterior mean values and variances of $\Lambda$
For a Gamma distribution with parameter $k$ and $\nu$, we have

$$f_\Lambda(\lambda) = \frac{\nu(\nu\lambda)^{k-1} e^{-\nu\lambda}}{\Gamma(k)} \Rightarrow \begin{matrix} E[\Lambda] = \dfrac{k}{\nu} \\ Var[\Lambda] = \dfrac{k}{\nu^2} \end{matrix}$$

Thus:

Table 9.2. Prior and posterior probabilities

| | $\nu'$ | $k$ | $E[\Lambda]$ | $Var[\Lambda]$ |
|---|---|---|---|---|
| Prior | $\nu' = 2$ | $k' = 1$ | $\frac{1}{2}$ | $\frac{1}{4}$ |
| Posterior | $\nu'' = 8$ | $k'' = 4$ | $\frac{1}{2}$ | $\frac{1}{16}$ |

Note that the evidence does not move the mean of the distribution but reduces considerably the variance, as indication of the fact that the posterior contains more information than the prior.

5. The 95% upper confidence bound and the 90% confidence interval for $\Lambda$

*Frequentist statistics*
The evidence $E = \{t_1 = 1, t_2 = 2.8, t_3 = 2.2\}$ constitutes the result of an uncensored test, but we can consider the test as a type 2 censoring test (test ends at the $r$-th item failure) with a total time to failure $T = 1 + 2.2 + 2.8 = 6$ and $r = 3$.

Using the Table for the normal standard variation we compute:

- The 95% upper bound:

$$\left(\frac{2T}{z_{95,2r}}\right)^{-1} = \left(\frac{2*6}{z_{95,6}}\right)^{-1} = \left(\frac{12}{12.6}\right)^{-1} = 1.05\,years^{-1}$$

where $Z_{100\varepsilon,\nu}$ denotes the upper $\varepsilon$ percentile of the chi-square distribution with $\nu$ degrees of freedom.

- The 90% confidence interval for $\Lambda$:

$$P\left(\left(\frac{2T}{z_{05,6}}\right)^{-1} \leq \lambda \leq \left(\frac{2T}{z_{95,6}}\right)^{-1}\right) = 0.9$$

$$P\left(\left(\frac{2*6}{z_{05,6}}\right)^{-1} \leq \lambda \leq \left(\frac{2*6}{z_{95,6}\,)}\right)^{-1}\right) = 0.9$$

$$P(0.1367 < \lambda \leq 1.05) = 0.9$$

*Bayesian statistics*

- The 95% upper bound:

The 95% upper confidence bound for $\Lambda$ is the value $\lambda_{95}$ such that,

$$P(\Lambda \leq \lambda_{95}) = \int_{-\infty}^{\lambda_{95}} \pi''(\lambda|E)d\lambda = \int_{-\infty}^{\lambda_{95}} \frac{\nu''[\nu''\lambda]^{k''-1}}{\Gamma(k'')} e^{-\nu''\lambda} d\lambda = 0.95$$

With $k'' = 4$ and $\nu'' = 8$

$$\frac{8^4}{6}\int_{-\infty}^{\lambda_{95}} \lambda^3 e^{-8\lambda} d\lambda = 0.95 \Rightarrow \lambda_{95} = 0.9676$$

- The 90% confidence interval for $\Lambda$:

Analogously the 90% confidence interval for $\Lambda$ is an interval $[\lambda_{05}, \lambda_{95}]$ such that,

$$P[\lambda_{0.05} < \Lambda < \lambda_{0.95} | E] = \int_{\lambda_{05}}^{\lambda_{95}} \pi''(\lambda|E) d\lambda = 0.90$$

We now compute $\lambda_{05}$:

$$P(\Lambda < \lambda_{05}) = \frac{8^4}{6} \int_{-\infty}^{\lambda_{05}} \lambda^3 e^{-8\lambda} d\lambda = 0.05 \Rightarrow \lambda_{05} = 0.1696$$

Thus, the 90% confidence interval for $\Lambda$ is [0.1696, 0.9676].
Note that the bayesian confidence interval is smaller than the frequentist confidence interval.

6. Censored test of 4 components
Now the evidence is regarded as the result of a Type I Censoring test (test ends at a fixed $t_0$). We have tested $m = 4$ items, with $n = 3$ failures before $t_0 = 2.8s$. Thus the total time to test $T$ is:

$$T = \sum_i t_i + (m - n) t_0 = (1 + 2.2 + 2.8) + (4 - 3) \cdot 2.8 = 8.8s$$

*Repeat 2*
Using Table 8.1 for conjugate distributions we find that the posterior distribution of $\Lambda$ is a Gamma distribution with parameter $\nu'' = \nu' + T = 10.8$ and $k'' = k' + n = 4$:

$$\pi''(\lambda|E) = \frac{10.8(10.8\lambda)^3 e^{-10.8\lambda}}{3!}$$

The posterior marginal distribution of $T$ is:

$$f(t|E) = \frac{10.8^4}{3!}\int_0^\infty \lambda^4 e^{-\lambda(10.8+t)}d\lambda = \frac{4\cdot 10.8^4}{(10.8+t)^5}$$

The posterior reliability of the item for a period of 1 year is:

$$R(1) = 1 - F(1) = 1 - \int_0^1 f(t)dt = F_m(t) = 1 - \int_0^1 \frac{54419}{(10.8+t)^5}dt = 0.702$$

*Repeat 3*

Table 9.3. Prior and posterior probabilities

| | $\nu'$ | $k$ | $E[\Lambda]$ | $Var[\Lambda]$ |
|---|---|---|---|---|
| Prior | $\nu' = 2$ | $k' = 1$ | $\frac{1}{2}$ | $\frac{1}{4}$ |
| Posterior | $\nu'' = 10.8$ | $k'' = 4$ | 0.37 | 0.034 |

Note that the evidence reduces considerably the variance, as indication of the fact that the posterior contains more information than the prior.

*Repeat 4*

*Frequentist statistics*

- The 95% upper bound:

$$\left(\frac{2T}{z_{95,2r+2}}\right)^{-1} = \left(\frac{2*8.8}{z_{95,8}}\right)^{-1} = \left(\frac{17.6}{15.5}\right)^{-1} = 0.8807\,years^{-1}$$

where $Z_{100\varepsilon,\nu}$ denotes the upper $\varepsilon$ percentile of the chi-square distribution with $\nu$ degrees of freedom.

- The 90% confidence interval for $\Lambda$:

$$P\left(\left(\frac{2T}{z_{05,2r}}\right)^{-1} \leq \lambda \leq \left(\frac{2T}{z_{95,2r+2}}\right)^{-1}\right) = 0.9$$

$$P\left(\left(\frac{2*8.8}{z_{05,6}}\right)^{-1} \leq \lambda \leq \left(\frac{2*8.8}{z_{95,8}\,)}\right)^{-1}\right) = 0.9$$

$$P(0.093 < \lambda \leq 1.05) = 0.9$$

*Bayesian statistics*

- The 95% upper bound:

The 95% upper confidence bound for $\Lambda$ is the value $\lambda_{95}$ such that:

$$P(\,\Lambda \leq \lambda_{95}\,) = \int_{-\infty}^{\lambda_{95}} \pi''(\,\lambda|E\,)d\lambda = \int_{-\infty}^{\lambda_{95}} \frac{\nu''[\,\nu''\lambda\,]^{k''-1}}{\Gamma(\,k''\,)} e^{-\nu''\lambda} d\lambda = 0.95$$

With $k'' = 4$ and $\nu'' = 10.8$

$$\frac{10.8^4}{6}\int_{-\infty}^{\lambda_{95}} \lambda^3 e^{-10.8\lambda} d\lambda = 0.95 \Rightarrow \lambda_{95} = 0.717$$

- The 90% confidence interval for $\Lambda$:

Analogously the 90% confidence interval for $\Lambda$ is an interval $[\,\lambda_{05}, \lambda_{95}\,]$ such that

$$P[\lambda_{05}<\Lambda<\lambda_{95}|E]=\int_{\lambda_{05}}^{\lambda_{95}} \pi''(\lambda|E)d\lambda = 0.90$$

We now compute $\lambda_{05}$:

$$P(\Lambda < \lambda_{05}) = \frac{10.8^4}{6} \int_0^{\lambda_{05}} \lambda^3 e^{-10.8\lambda} d\lambda = 0.05 \Rightarrow \lambda_{05}=0.126$$

Thus, the 90% confidence interval for $\Lambda$ is [0.126, 0.717].
Note that the bayesian confidence interval is smaller than the frequentist confidence interval.

## 9.6 Pile failure probability estimation

Consider the problem in which the probability of pile failure at a load of 300 tons is of concern; this time, however, assume that the probability $p$ is a continuous random variable. If there is no prior factual information on $p$, a uniform prior distribution $P'(p)$ may be assumed.
1. Based on the single failure test result revise the probability $P'(p)$ and obtain the new estimate for $p$.
2. If a sequence of n piles were tested, out of which r piles failed at loads less than the maximum test load, revise the probability distribution $P(p)$ and obtain the new estimate for $p$ [Suggestion: integrate by parts repeatedly].
3. Discuss the results.

**Solution**

1. Probability of failure at 300 tons
The prior uniform distribution may be written as:

$$f'(p) = 1.0 \qquad 0 \le p \le 1$$

On the basis of a single test, the likelihood function is the probability of the event
$E$ = [capacity of test pile less than 300 tons]
which is simply $p$. Hence the posterior distribution of $p$ is:

$$f''(p) = \frac{p \cdot f'(p)}{\int_0^1 p \cdot f'(p) dp} = \frac{p}{\frac{1}{2}} = 2p$$

Thus

$$f''(p) = 2p \qquad 0 \leq p \leq 1$$

The Bayesian estimate of $p$ is:

$$\hat{p}'' = E[\, p \,/\, E \,] = \int_0^1 p \cdot 2p \, dp = 0.667$$

2. Revised probability after one test failure
If a sequence of $n$ piles were tested, out of which $r$ piles failed at loads less than the maximum test load, the likelihood function is the probability of observing $r$ failures among the $n$ piles tested. If the failure probability of each pile is $p$ and statistical independence is assumed between piles, the likelihood function would be:

$$L(\, p \,) = B(r \,/\, n, p) = \binom{n}{r} p^r (\, 1 - p \,)^{n-r}$$

Then, the posterior distribution of $p$ becomes

$$f''(p \mid E) = \frac{f'(p)L(p)}{\int_{-\infty}^{\infty} f'(p)L(p)dp} = \frac{1 \cdot \binom{n}{r} p^r (1-p)^{n-r}}{\int_0^1 1 \cdot \binom{n}{r} p^r (1-p)^{n-r} dp}$$

$$= \frac{p^r (1-p)^{n-r}}{\int_0^1 p^r (1-p)^{n-r} dp}$$

Integrating by parts we obtain

$$f''(p / E) = \frac{p^r (1-p)^{n-r}}{\int_0^1 \frac{(1-p)^{n-r+1}}{n-r+1} r p^{r-1} dp}$$

Integrating by parts again we obtain:

$$f''(p / E) = \frac{p^r (1-p)^{n-r}}{\int_0^1 \frac{(1-p)^{n-r+2}}{(n-r+1)(n-r+2)} r(r-1) p^{r-2} dp}$$

Repeated integration-by-parts of the above integral yields:

$$f''(p \mid x) = \binom{n}{r} (n+1) p^r (1-p)^{n-r}$$

We can now look for the new bayesian estimate of $p$:

$$\hat{p}'' = \langle p \rangle = \int_0^1 p P''(p) dp = \int_0^1 \binom{n}{r} (n+1) p^{r+1} (1-p)^{n-r} dp$$

$$= \binom{n}{r} (n+1) \int_0^1 p^{r+1} (1-p)^{n-r} dp$$

$$\hat{p}'' = \frac{r+1}{n+2}$$

From this result, we may observe that as the number of tests $n$ increases (with the ratio $\frac{r}{n}$ remaining constant), the bayesian estimate for $p$ approaches that of the classical estimate; that is,

$$\frac{r+1}{n+2} \to \frac{r}{n} \qquad \text{for } n \to \infty$$

**9.7 Failure rate estimation**

Consider a plant which has a specified number of identical and independent valves with constant failure rate $\lambda$, where $\lambda$ represents a realization of a random variable $\Lambda$ with the Gamma prior density

$$f'_\Lambda(\lambda) = \frac{\beta^{\alpha'}}{\Gamma(\alpha')} \lambda^{\alpha'-1} e^{-\beta'\lambda} \qquad \text{for} \qquad \begin{matrix} \alpha' > 0 \\ \beta' > 0 \\ \lambda > 0 \end{matrix}$$

The parameters $\alpha'$ and $\beta'$ of the prior distribution are usually "estimated" based on prior experience with the same type of valves, combined with information gained from various reliability data sources. In our case, $\alpha' = 3$, $\beta' = 10^4$. When a valve fails, it will be replaced with a valve of the same type. The associated downtime is considered to be negligible. If the number of valve failures during an accumulated service time $t = 5 \cdot 10^3$ hours is $n = 2$:

1. Find the posterior density of $\Lambda$.
2. What are the prior and posterior mean values and variances of $\Lambda$.
3. Find the bayesian estimate of $\Lambda$.
4. Define the 90% confidence interval for $\Lambda$ and find it.
5. Find an estimator of $\Lambda$ using frequentist statistics and its 90% confidence interval.

**Solution**

1. Posterior density

Valve failures are assumed to occur according to a homogeneous Poisson process with intensity. The number of valve failures during an accumulated time $t$ in service thus has the distribution:

$$P(N(t) = n \mid \Lambda = \lambda) = \Pi(n, (0,t) \mid \lambda) = \frac{(\lambda t)^n}{n!} e^{-\lambda t} \quad \text{for} \quad n = 0,1,...$$

The posterior density of $\Lambda$ is:

$$f''_\Lambda(\lambda) = \frac{f'_\Lambda(\lambda) \cdot P(N(t) = n \mid \Lambda = \lambda)}{\int_0^\infty f'_\Lambda(\lambda) \cdot P(N(t) = n \mid \Lambda = \lambda) d\lambda}$$

where

$$\int_0^{+\infty} f'_\Lambda(\lambda) \cdot P(N(t) = n \mid \Lambda = \lambda) d\lambda =$$

$$\int_0^{+\infty} \frac{(\lambda t)^n}{n!} e^{-\lambda t} \cdot \frac{\beta'^{\alpha'}}{\Gamma(\alpha')} \lambda^{\alpha'-1} e^{-\beta'\lambda} d\lambda =$$

$$= \frac{\beta'^{\alpha'} t^n}{\Gamma(\alpha') n!} \int_0^{+\infty} \lambda^{\alpha'+n-1} e^{-(\beta'+t)\lambda} d\lambda = \frac{\beta'^{\alpha'} t^n}{\Gamma(\alpha') n!} \frac{\Gamma(n+\alpha')}{(\beta'+t)^{n+\alpha'}}$$

so that

$$f''_{\Lambda}(\lambda \mid n) = \frac{(\beta' + t)^{\alpha'+n}}{\Gamma(\alpha' + n)} \lambda^{\alpha'+n-1} e^{-(\beta'+t)\lambda}$$

which is recognized as the gamma distribution with parameters $(\alpha' + n)$ and $(\beta' + t)$. Hence, we have confirmed that the family of Gamma distributions $\wp$ is conjugate to the family of Poisson distributions $\Im$:

$$f'_{\Lambda}(\lambda) \in \wp \text{ and } P(N(t) = n \mid \Lambda = \lambda) \in \Im \Rightarrow f''_{\Lambda}(\lambda \mid n) \in \wp$$

2. Prior and posterior mean values and variance of $\Lambda$
For a Gamma distribution with parameter $k$ and $\nu$, we have

$$f_{\Lambda}(\lambda) = \frac{\nu(\nu\lambda)^{k-1} e^{-\nu\lambda}}{\Gamma(k)} \Rightarrow \begin{array}{l} E(\lambda) = \dfrac{k}{\nu} \\ Var(\lambda) = \dfrac{k}{\nu^2} \end{array}$$

Thus:

Table 9.4. Prior and posterior probabilities

| | $\nu$ | $k$ | $E(\lambda)$ | $Var(\lambda)$ |
|---|---|---|---|---|
| Prior | $\beta' = 10^4$ | $\alpha' = 3$ | $3 \cdot 10^{-4}$ | $3 \cdot 10^{-8}$ |
| Posterior | $\beta'' = \beta' + t$ $= 1.5 \cdot 10^4$ | $\alpha'' = \alpha' + n = 5$ | $3.33 \cdot 10^{-4}$ | $2.22 \cdot 10^{-8}$ |

Note that the evidence reduces the variance, as indication of the fact that the posterior contains more information than the prior.

3. Bayesian estimate of $\Lambda$

The Bayesian estimate of $\Lambda$, when the evidence $N(t)=n$ is accounted for, is

$$\hat{\lambda}'' = E(\Lambda \mid N(t)=n) = \frac{\alpha''}{\beta''} = \frac{\alpha'+n}{\beta'+t} = 3.33 \cdot 10^{-4}\, h^{-1}$$

4. 90% Confidence interval for $\Lambda$

The confidence interval for $\lambda$, at level $1-2\varepsilon$, is an interval $[a(N,t), b(N,t)]$, such that, given the data $(N,t)$,

$$P[a(N,t) < \lambda < b(N,t) | N,t] = \int_{a(N,t)}^{b(N,t)} f''_{\Lambda}(\lambda \mid N,t) d\lambda = 1-2\varepsilon$$

If we define a variable $Z = 2(\beta + t)\Lambda$, we must have:

$$f''_{\Lambda}(\lambda) d\lambda = g''_{Z}(z) dz$$

and so:

$$g''_{Z}(z) = \frac{1}{2^{\alpha'+n}\Gamma(\alpha'+n)} z^{\alpha'+n-1} e^{-\frac{z}{2}} \qquad for \qquad z > 0$$

which is recognized as the chi-square distribution with $2(\alpha'+n)$ degrees of freedom (e.g., see Dudewicz and Mishra, 1998, p 140). The $1-2\varepsilon = 0.9$ confidence interval for the failure rate is then easily obtained from the chi-square Table as:

$$P(z_{5,10} \le z < z_{95,10}) = \int_{z_{5,10}}^{z_{95,10}} g''_{Z}(z) dz = \int_{z_{5,10}/2(\beta+t)}^{z_{95,10}/2(\beta+t)} f''_{\Lambda}(\lambda) d\lambda =$$

$$= P\left( \frac{z_{5,10}}{2(\beta+t)} < \lambda < \frac{z_{95,10}}{2(\beta+t)} \mid N(t) = 2 \right) = 0.9$$

where $z_{\varepsilon,\upsilon}$ denotes the ε percentile of the chi-square distribution with ν degree of freedom.

$$P\left(1.31\cdot 10^{-4} < \lambda < 6.1\cdot 10^{-4} \mid N(t)=2\right) = 0.9$$

5. Frequentist estimate of Λ and 90% confidence interval
Under the assumptions made failures will occur according to a Poisson process and the number of failures $N$ observed during the period t will have a Poisson distribution with parameter $\lambda t$. Using the method of momentum we can find an unbiased estimator of the failure rate $\lambda$:

$$\hat{\lambda}_{MOM} = \frac{N}{t} = \frac{2}{5\cdot 10^{3}} = 4\cdot 10^{-4}\,h^{-1}$$

and the $1-2\varepsilon = 0.9$ confidence interval for $\lambda$ is given by:

$$\left(\frac{1}{2t}z_{5,2N}, \frac{1}{2t}z_{95,2(N+1)}\right)$$

where $z_{100\varepsilon,\upsilon}$ denotes the upper ε percentile of the chi-square distribution with ν degrees of freedom. In this case

$$\begin{matrix} z_{100\varepsilon,2N} = z_{5,4} = 0.711 \\ z_{100\cdot(1-\varepsilon),2N+2} = z_{95,6} = 12.61 \end{matrix} \Rightarrow P\left(1.42\cdot 10^{-4} < \Lambda < 25\cdot 10^{-4}\right) = 0.9$$

## 9.8 Bayesian estimation

A statistician is interested in the mean μ of a normal population. The variance of the population is known to be $\sigma^2 = 72$. The prior distribution of the mean $\mu$ is normal with mean $\mu' = 800$ and variance $(\sigma')^2 = 12$.

1. How large a sample would be needed to guarantee that?
2. And no larger than 0.1?

**Solution**

1. Sample dimension for variance of the posterior distribution no larger than 1

The sampling distribution is normal. The variance of the population is $\sigma^2 = 72$. The parameter to be estimated is μ.
By exploiting the formulas for the conjugate distributions

$$\text{Sampling distribution} = \frac{1}{\sigma\sqrt{2\pi}} e^{-\frac{(z-\mu)^2}{2\sigma^2}}$$

$$\text{Prior distribution} = \frac{1}{\sigma'\sqrt{2\pi}} e^{-\frac{(\mu-\mu')^2}{2(\sigma')^2}}$$

$$\text{Posterior distribution: } \frac{1}{\sigma''\sqrt{2\pi}} e^{-\frac{(\mu-\mu'')^2}{2(\sigma'')^2}}$$

The variance of the posterior is (Table A-1 in Appendix):

$$\sigma'' = \frac{(\sigma')^2\sigma^2}{\sigma^2 + n(\sigma')^2} = \frac{12 \cdot 72}{72 + n \cdot 12} = \frac{864}{72 + n \cdot 12}$$

To have $(\sigma'')^2 \leq 1$, $n \geq 66$

2. Sample dimension for variance of the posterior distribution no larger than 0.1

To have $(\sigma'')^2 \leq 0.1$, $n \geq 314$

## 9.9 Defective rivets on airplane wing

The number of defective rivets, D, on an airplane wing can be assumed to have a Poisson distribution with parameter $\lambda$, i.e.,

$$P(D = k) = \frac{\lambda^k e^{-\lambda}}{k!}, \quad k = 1,2,...$$

A random sample of n wings is observed.
1. What is $\hat{\lambda}$, the maximum likelihood estimator of $\lambda$?
2. Is this estimator unbiased?
3. Find the method-of-moments estimator of $\lambda$.
4. Find the bayes estimator of $\lambda$ assuming a uniform prior density function.

### Solution

Let $r_i$ be the number of defective rivets found in the i-th wing, $i = 1,2,...,n$ i = 1, 2, …, n.
1. Maximum likelihood estimator
The likelihood function of the evidence is:

$$L(\lambda) = \prod_{i=1}^{n} \frac{\lambda^{r_i} e^{-\lambda}}{r_i!} = \frac{\lambda^{\left(\sum_{i=1}^{n} r_i\right)} e^{-n\lambda}}{\prod_{i=1}^{n} r_i!}$$

The maximum of the likelihood function $L(\lambda)$ is the maximum likelihood estimator of $\lambda$. To simplify the calculation to get the maximum, we take the logarithm of the function $L(\lambda)$ in the previous equation:

$$ln(\,L(\,\lambda\,)) = \left(\sum_{i=1}^{n} r_i\right) \cdot ln\,\lambda - n\lambda - \sum_{i=1}^{n} ln(\,r_i!\,)$$

Now, we differentiate the expression above with respect to $\lambda$ and set this derivative equal to zero to find the value of $\lambda$ which maximizes $\ln(L(\lambda))$:

$$\frac{d(ln\,L(\,\lambda\,))}{d\lambda} = \frac{\sum_{i}^{n} r_i}{\lambda} - n \;\Rightarrow\; \hat{\lambda}_{MLE} = \frac{\sum_{i}^{n} r_i}{n} = \frac{r}{n}$$

where $r = \sum_{i}^{n} r_i$ is the total number of defective rivets.

2. Unbiased estimator?

An estimator of the parameter is said to be unbiased if its expected value is equal to the parameter itself. The expected value of $\hat{\lambda}_{MLE}$, $E[\hat{\lambda}]$, is

$$E[\,\hat{\lambda}_{MLE}\,] = \frac{E\left[\sum_{i}^{n} r_i\right]}{n} = \frac{\sum_{i=1}^{n} E[\,r_i\,]}{n} = \frac{n\lambda}{n} = \lambda\,.$$

Therefore, $\hat{\lambda}$, the maximum likelihood estimator of $\lambda$, is unbiased.

3. Method-of-moments estimator

The sample first moment, i.e. the sample mean, of the random variable D, can be utilized to estimate the mean of the underlying Poisson distribution, i.e. the parameter $\lambda$:

$$\hat{\lambda}_{MOM} = \frac{\sum_{i}^{n} r_i}{n}\,.$$

Therefore, the methods of moments yields the same estimator as the method of moments.

4. Bayes estimator

The Bayes estimator, $\hat{\lambda}_B$, is the mean of the posterior distribution. The prior uniform distribution can be interpreted as a particular gamma distribution with parameters $k'=1$ and $\nu'=0$. Since the gamma distribution and the Poisson are conjugated, the posterior distribution will be also gamma and its parameters can be obtained as a function of the evidence of $x=\sum_{i=1}^{n} r_i$ defective rivets found in $t=n$ wings:

$$k''=k'+x=1+\sum_{i=1}^{n} r_i=1+r$$

$$\nu''=\nu'+t=n$$

Thus, the expected value of the posterior gamma distribution, i.e. the Bayes estimator , $\hat{\lambda}_B$, is:

$$\hat{\lambda}_B=\frac{k''}{\nu''}=\frac{r+1}{n}.$$

Note that this estimator is different from the previous two, particularly for small sample sizes $n$.

**9.10 Start on demand failures**

The reactor safety study has assessed the range of the frequency $q$ of failure of pumps to start on demand as:

$$q \in [3\cdot 10^{-4}, 3\cdot 10^{-3}]$$

The data from a particular plant are: 3 failures to start in 500 tests.

1. What is the posterior distribution?

2. What are the posterior 5th, 50th and 95th percentiles? Compare with the corresponding values of the prior distribution.
3. Compare the mean and variances of the prior and posterior distribution.
4. Find the lognormal distribution with mean and variance values equal to those of the posterior distribution. How good an approximation is it?

**Solution**

1. Posterior distribution

We want to develop site specific failure data combining the generic assessed range and the specific plant data, through Bayes' theorem. The assessed range of frequency of failure is $[3 \cdot 10^{-4}\ 3 \cdot 10^{-3}]$. A practical assumption is that the two extrema represent the 5-th and 95-th percentiles of the distribution of $q$. The site specific evidence is 3 failures out of 500 tests.

Let us first assume that the a priori distribution of the probability of failure per demand $q$ is lognormal. Then, we obtain the values of the parameters of the distribution, $\mu_z$ and $\sigma_z$, from the data of the problem:

$$\begin{cases} q_{0.95} = 3 \cdot 10^{-3} = e^{\mu_z + z_{95}\sigma_z} = e^{\mu_z + 1.645\sigma_z} \\ q_{0.05} = 3 \cdot 10^{-4} = e^{\mu_z + z_{95}\sigma_z} = e^{\mu_z - 1.645\sigma_z} \end{cases} \Rightarrow$$

$$\begin{cases} \mu_z + 1.645 \cdot \sigma_z = \ln(3 \cdot 10^{-3}) = -5.809 \\ \mu_z - 1.645 \cdot \sigma_z = \ln(3 \cdot 10^{-4}) = -8.112 \end{cases} \Rightarrow \begin{cases} \mu_z = -6.96 \\ \sigma_z = 0.70 \end{cases}$$

Therefore, the prior distribution of the failure rate $q$ is $P'(q) = lN(-6.96, 0.70)$.

The binomial distribution can be utilized to find the probability of the evidence E of 3 failures out of 500 tests, given that $q$ is the probability of failure in a single test. Correspondingly, the likelihood function, $L(E|q)$, reads:

$$L(E|q) = \binom{500}{3} \cdot q^3 \cdot (1-q)^{497}$$

Using the Bayes' theorem, the posterior distribution, $P''(q)$, is obtained as follows:

$$P''(q) = \frac{P'(q) \cdot L(E|q)}{\int_0^{\infty} P'(q) \cdot L(E|q)}.$$

Unfortunately, the binomial and lognormal distributions are not conjugate families, so that a closed form analytical solution does not exist. We must resort to numerical techniques, in particular the discretization of the distribution (Figure 9).

$$\Delta A(q_i) = \frac{P'(q_i) + P'(q_{i+1})}{2} \cdot \Delta q$$

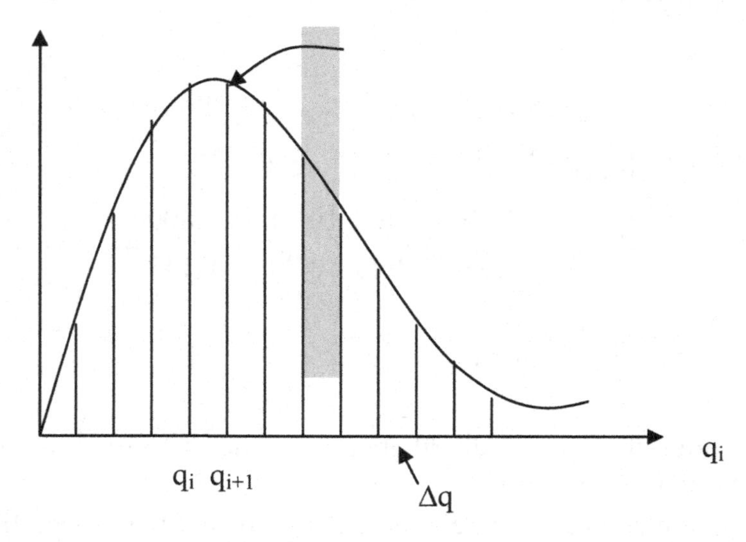


Figure 9.4. The technique of the discretization of the prior distribution

The q-axis is divided into *n* intervals of equal amplitude $\Delta q = 10^{-4}$ (Figure 9.4).
For each point $q_i$ we evaluate the prior distribution:

$$P'(q_i) = \frac{1}{q_i \cdot \sigma_z \cdot \sqrt{2\pi}} \cdot e^{-\frac{1}{2}\left(\frac{-\mu_z + \ln q_i}{\sigma_z}\right)^2}$$

Figure 9.5 reports the prior lognormal distribution. As discrete value representing the generic interval $(q_i, q_{i+1})$ we assign to $q_{i+1/2}$ the value $P'(q_{i+1/2})$ which is the average of $P'(q_i)$ and $P'(q_{i+1})$:

$$P'(q_{i+1/2}) = \frac{P'(q_i) + P'(q_{i+1})}{2}$$

The incremental probability area $\Delta A(q_i)$ is then approximated by a rectangle of height equal to $P'(q_{i+1/2})$ and width equal to $\Delta q$:

$$\Delta A(q_i) = P'(q_{i+1/2}) \cdot \Delta q = \frac{P'(q_i) + P'(q_{i+1})}{2} \cdot \Delta q \, .$$

At the j-th point, the sum of $\Delta A(q_i)$ up to that point gives the cumulative distribution at $q_j$:

$$F(q_j) = \sum_{i=1}^{j+1} \Delta A(q_i) \, .$$

The likelihood function given by the binomial distribution is evaluated at the midpoint of each interval, $q_{i+1/2}$:

$$L(E|q_{i+1/2}) = \binom{500}{3} \cdot q_{i+1/2}^{3} \cdot (1 - q_{i+1/2})^{497}$$

The integral at the denominator of the Bayes' theorem can be then approximated as follows:

$$\int_0^\infty P'(q) \cdot L(E|q) = \sum_{i=1}^{n} \Delta A(q_i) \cdot L(E|q_{i+1/2})$$

By so doing, we numerically evaluate the posterior distribution $P''(q_{i+1/2})$ in each $q_{i+1/2}$:

$$P''(q_{i+1/2}) = \frac{P'(q_{i+1/2}) \cdot L(E|q_{i+1/2})}{\sum_{i=1}^{n} \Delta A(q_i) \cdot L(E|q_{i+1/2})}.$$

The behaviour of $P''(q_{i+1/2})$ is reported in Figure 9.6.

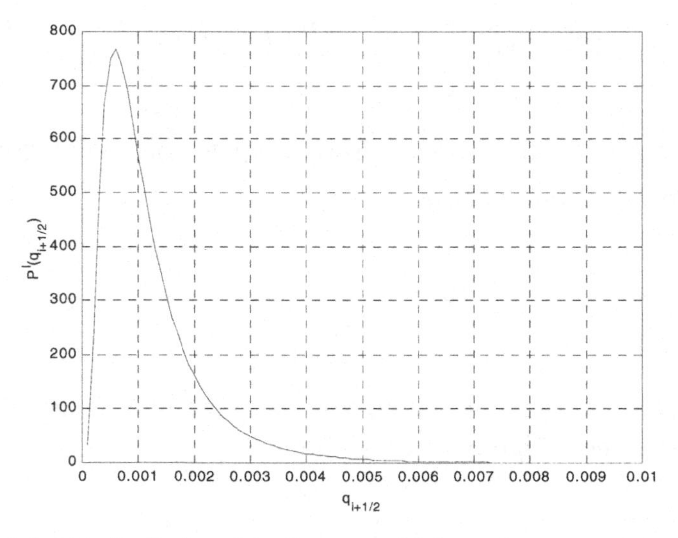


Figure 9.5. Prior distribution of the frequency of failures of pumps to start on demand

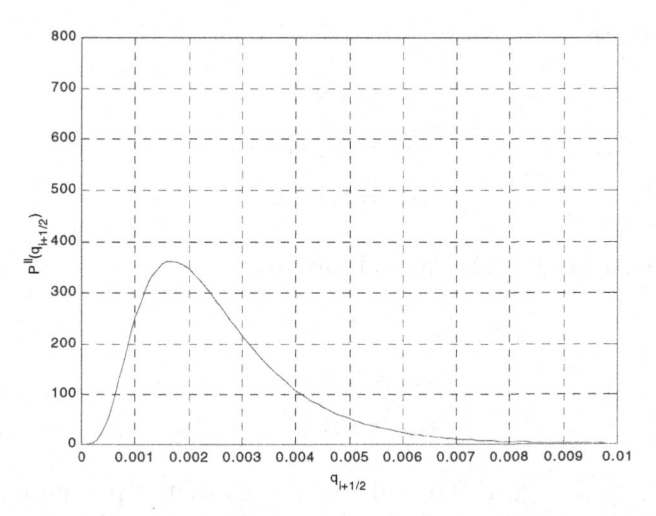


Figure 9.6. Posterior distribution of the frequency of failures of pumps to start on demand

2. Posterior percentiles

Figure 9.7 shows the cumulative distribution of the posterior probability function. The $\alpha$-percentile are found as those values $q_\alpha$ at which the cumulative probability distribution is $\alpha\%$.

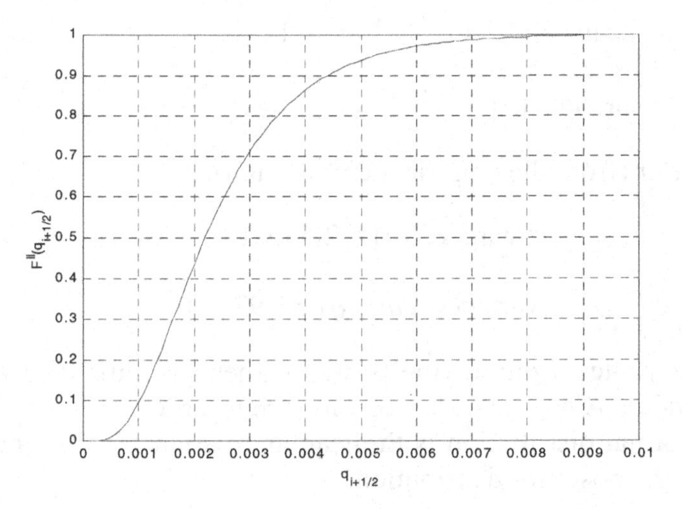


Figure 9.7. Posterior cumulative distribution of the frequency of failures of pumps to start

By interpolation, we find:

$$q_{05} = 8.8 \cdot 10^{-4}$$
$$q_{50} = 2.3 \cdot 10^{-3}$$
$$q_{95} = 5.3 \cdot 10^{-3}$$

For comparison, the prior percentiles are:

$$q_{05} = 3 \cdot 10^{-4}$$
$$q_{50} = 9.49 \cdot 10^{-4}$$
$$q_{95} = 3 \cdot 10^{-3}$$

Figure 9.5 and Figure 9.6 and the values of the percentiles show that the posterior is shifted towards higher values of $q$, thus denoting a non conservative a priori information.

3. Comparison between the mean and variances of the prior and posterior distribution
A priori:

$$\text{mean: } E'[q] = e^{-\mu_z + \frac{\sigma_z^2}{2}} = 1.212 \cdot 10^{-3}$$

$$\text{var: } Var'[q] = e^{-2\mu_z + \sigma_z^2}(e^{-\mu\sigma_z^2} - 1) = 9.258 \cdot 10^{-7}$$

A posterior (from the empirical distribution):

$$\text{mean: } E''[q] \cong 2.546 \cdot 10^{-3}$$

$$\text{variance } Var''[q] \cong 1.97 \cdot 10^{-6}$$

So, as expected from 1., the posterior mean is shifted towards a higher value. Also, the posterior variance is larger.
4. Lognormal distribution with mean and variance values equal to those of the posterior distribution.

The parameters of the approximating lognormal with the same mean and variance of the posterior are found as follows:

$$\begin{cases} E''[q] = 2.546 \cdot 10^{-3} = e^{\mu_z + \frac{\sigma_z^2}{2}} \\ Var''[q] = 1.97 \cdot 10^{-6} = e^{2\mu_z + \sigma_z^2}(e^{\sigma_x^2} - 1) \end{cases} \Rightarrow \begin{cases} \mu_z = -6.106 \\ \sigma_z = 0.515 \end{cases}$$

The two curves, the posterior $P''(q)$ and the approximating lognormal are reported in Figure 9.8. To check the goodness of the approximation, we find the percentiles of the approximating distribution:

$$q_{0.05} = e^{\mu_z - 1.645\sigma_z} = e^{-6.106 - 1.645 \cdot 0.515} = 9.56 \cdot 10^{-4}$$
$$q_{0.50} = e^{\mu_z} = e^{-6.106} = 2.23 \cdot 10^{-3}$$
$$q_{0.95} = e^{\mu_z + 1.645\sigma_z} = e^{-6.106 + 1.645 \cdot 0.515} = 5.2 \cdot 10^{-3}.$$

The agreement with the posterior percentiles suggest that the approximation is satisfactory.

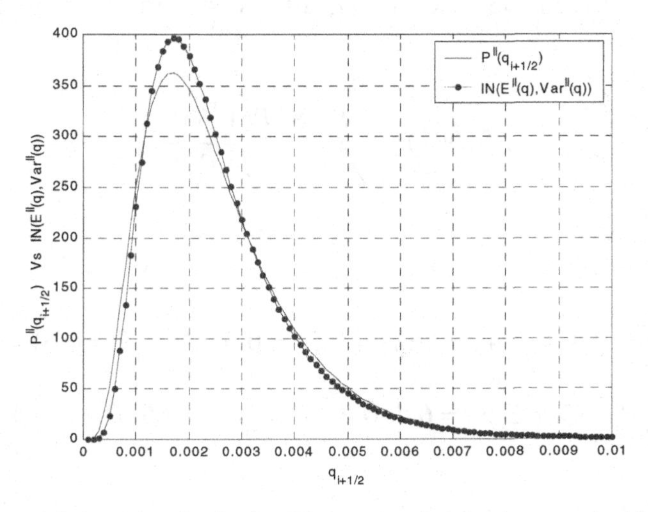


Figure 9.8. Posterior distribution Vs the approximating lognormal with same mean and variance.

## 9.11 Coin tossing

You feel that the frequency of heads on tossing a particular coin is either 0.4, 0.5 or 0.6. Your prior probabilities are $P'(0.4) = 0.1$, $P'(0.5) = 0.7$ and $P'(0.6) = 0.2$. You toss the coin just once.

1. Apply the Bayes' theorem given that the toss results in heads.
2. Repeat given that it results in tails.

In each case, consider the denominator of Bayes' theorem and interpret it.

**Solution**

1. Bayes' theorem application given that the toss results in heads

$$P'(\vartheta = 0.4) = 0.1$$
$$P'(\vartheta = 0.5) = 0.7$$
$$P'(\vartheta = 0.6) = 0.2$$

Apply the Bayes' theorem to revise the probabilities of the frequencies given the evidence $E$ of a result in heads in a single toss:

$$P''(\vartheta|\mathrm{E}) = \frac{P'(\vartheta)P(\mathrm{E}|\vartheta)}{\sum_{\vartheta} P'(\vartheta)P(\mathrm{E}|\vartheta)}$$

In our case,

$$P(E|\vartheta = 0.4) = 0.4,\ P(E|\vartheta = 0.5) = 0.5,\ P(E|\vartheta = 0.6) = 0.6.$$

$$\sum_{\vartheta} P'(\vartheta)P(E|\vartheta) = 0.1 \cdot 0.4 + 0.5 \cdot 0.7 + 0.6 \cdot 0.2 = 0.51$$

Thus, the posterior distribution is given by:

$$P''(\vartheta = 0.4 \mid E) = \frac{0.1 \cdot 0.4}{0.51} = 0.0784,$$

$$P''(\vartheta = 0.5 \mid E) = \frac{0.5 \cdot 0.7}{0.51} = 0.6863,$$

$$P''(\vartheta = 0.6 \mid E) = \frac{0.6 \cdot 0.2}{0.51} = 0.2353.$$

$\sum_{\vartheta} P(E|\vartheta)P'(\vartheta) = P(E)$ can be interpreted as the probability of a result in heads in a single trial based on our prior knowledge

2. Bayes' theorem application given that the toss results in tail

$$P(E|\vartheta = 0.4) = 1 - 0.4 = 0.6$$
$$P(E|\vartheta = 0.5) = 1 - 0.5 = 0.5$$
$$P(E|\vartheta = 0.6) = 1 - 0.6 = 0.4.$$

$$\sum_{\vartheta} P'(\vartheta)P(E|\vartheta) = 0.1 \cdot 0.6 + 0.7 \cdot 0.5 + 0.2 \cdot 0.4 = 0.49$$

$$P'(\vartheta = 0.4|E) = \frac{0.1 \cdot 0.6}{0.49} = 0.1224$$

$$P'(\vartheta = 0.5|E) = \frac{0.7 \cdot 0.5}{0.49} = 0.7143$$

$$P'(\vartheta = 0.6|E) = \frac{0.2 \cdot 0.4}{0.49} = 0.1633.$$

A similar interpretation to that made in case 1. can be drawn for the denominator $\sum_{\vartheta} P(E|\vartheta)P'(\vartheta) = P(E)$ in this case.

## 9.12 Failure probability estimation

A component has probability of success (per demand) equal to $p$. The prior distribution of $p$ is the beta distribution with mean and variance equal to 0.4286 and 0.0306, respectively. In a test involving 10 components, 8 components survive.

1. Find the posterior distribution of $p$.
2. Find the Bayes and the maximum likelihood estimates of $p$.
3. What is the probability that, if we repeat the experiment with 10 components, again 8 will survive?

### Solution

1. Posterior distribution of p

The distribution of the random variable $K$ = [number of successes in $n$ trial], given any particular value of $p$, is a binomial distribution. The likelihood $L(p|E)$ of the evidence E of $k$ = 8 components surviving out of $n$ = 10 components tested is:

$$L(p\,/\,E) = B(k\,/\,n,p) = \binom{n}{k} p^k (1-p)^{n-k} = \binom{10}{8} p^8 (1-p)^2$$

The prior distribution $f'(p)$ is a beta distribution with

$$\begin{cases} \overline{p} = 0.4236 \\ \sigma_p^2 = 0.0326 \end{cases}$$

First of all we find the parameters $q'$ and $r'$ of the beta distribution.

$$\begin{cases} \overline{p} = \dfrac{q'}{q'+r'} \\ \sigma_p^2 = \dfrac{q'r'}{(q'+r')^2(q'+r'+1)} \end{cases} \Rightarrow \begin{matrix} q' = 3 \\ r' = 4 \end{matrix}$$

We note that binomial and beta distribution are conjugate distributions, so the posterior distribution is also beta with parameters

$$\begin{aligned} q'' &= q' + k = 11 \\ r'' &= r' + n - k = 6 \end{aligned} \Rightarrow f''(p) = \frac{\Gamma(11+6)}{\Gamma(11)\Gamma(6)} p^{11-1}(1-p)^{6-1}$$

$$= 48048 \cdot p^{10}(1-p)^5$$

2. Bayesian and maximum likelihood estimates of $p$

The bayesian estimator of $p$ is the mean of the posterior distribution:

$$\hat{p}'' = \int_0^\infty p \cdot f''(p)\,dp = \frac{q''}{q''+r''} = \frac{11}{11+6} = 0.647$$

The likelihood function of the evidence is:

$$L(p \mid E) = \binom{10}{8} p^8 (1-p)^2$$

The maximum of the likelihood function $L(p|E)$ is the maximum likelihood estimator of $p$. We differentiate the expression above with respect to $p$ and set this derivative equal to zero to find the value of p which maximizes $L(\varepsilon|p)$:

$$\frac{dL}{dp} = \binom{10}{8}\left(8p^7(1-p)^2 + p^8 \cdot 2(1-p)\right) = 0 \rightarrow \hat{p} = 0.8$$

3. Probability of 8 components surviving the test out of 10

P(8 out of 10 will survive)

$$= \int_0^1 P[8\, out\, of\, 10 \mid p] \cdot f''(p)\, dp$$

$$\int_0^1 \binom{10}{8} p^8 (1-p)^2 \cdot 48048 p^{10} (1-p)^5\, dp = 0.173$$

**9.13 Life testing**

During life testing of a given component kind, one unit fails at 48 hours, another at 64 hours, and the other six units are tested without failing until hour 80. Furthermore, these is certain prior knowledge on the failure rate $\lambda$ of the component i.e. that $\lambda_0 = 0.4 \cdot 10^{-4} h^{-1}$ and $\sigma_{\lambda_0} = 0.1 \cdot 10^{-4} h^{-1}$.

1. Find the bayesan estimator of $\lambda$.
2. Find the 95% upper confidence bound for $\lambda$.

**Solution**

1. Bayesian estimate of λ

Firstly, we calculate the total time on test:

$$t = \sum_i t_i + (m-n)t_0$$

where

$t_i$ = failure time of unit $i$

$m$ =number of tested units=8

$n$ =number of units that have a failure during the test time=2

$t_0$ =test time=80 h

$$t = 48 + 64 + 6 \cdot 80 = 592h$$

We then assume a prior gamma distribution, because conjugate to the exponential, and estimate the value of the parameters $k$, $\nu$ of the prior gamma distribution, from the known values of $\lambda_0$ and $\sigma_{\lambda_0}$:

$$\begin{cases} E(\lambda) = \dfrac{k'}{\nu'} = \lambda_0 \\ Var(\lambda) = \dfrac{k'}{\nu'^2} = (\sigma_{\lambda_0})^2 \end{cases} \Rightarrow \begin{cases} k' = 16 \\ \nu' = 0.4 \cdot 10^6 \end{cases}$$

Given that prior (Gamma) and likelihood (exponential) distributions are conjugate, the posterior distribution is also a Gamma distribution with parameters

$$\begin{cases} k'' = k' + n = 16 + 2 = 18 \\ \nu'' = \nu' + t = 4 \cdot 10^5 + 592 = 400592 \end{cases}$$

and the bayesian estimator of $\lambda$ is

$$\hat{\lambda}'' = \frac{k''}{\nu'''} = 4.49 \cdot 10^{-5}\ \text{h}^{-1}$$

2. 95% upper confidence bound for λ

A one-sided upper confidence limit 1-ε for the failure rate λ is obtained by:

$$P(\lambda < \frac{z_{\varepsilon,2(k+n)}}{2(\nu + t)}) = 1 - \varepsilon$$

Thus:

$$z_{10,36} = 47.212 \Rightarrow P(\lambda < 0.589 \cdot 10^{-4}\ f/h) = 0.9$$

## 9.14 Defective items in a manufacturing process

A production manager is concerned about the proportion of defective items produced by a certain manufacturing process. From past experience he fells that $p$, the proportion of defectives, can take only four different values, 0.01, 0.05, 0.10, 0.25, based on the type of malfunction that has occurred.
Suppose now that the production manager has some information concerning $p$. This information can be summarized in terms of the following degree-of-belief probabilities for the four possible values of $p$ (ses Figure 9.9):

$$P'(p = 0.01) = 0.60$$
$$P'(p = 0.05) = 0.30$$
$$P'(p = 0.10) = 0.08$$
$$P'(p = 0.25) = 0.02\,.$$

In words, the probability is 0.60 that 1 percent of the items are defective, the probability is 0.30 that 5 percent of the items are defective, and so on.
Moreover, the production manager assumes that the process can be thought of as a Bernoulli process, with the assumptions of stationarity and independence appearing reasonable. That is, the probability that any one item is defective remains constant for all items produced and is independent of the past history of defectives from the process.

1. A sample of 5 items is taken from the production process, and 1 of the 5 is found to be defective. How can this information be combined with the prior information? Use a graph to show how the a priori distribution is modified in lights of the evidence.
2. Suppose now that a new sample of 5 items is taken and that 2 defectives are found. Revise the information inferred in (1.).
3. If instead of performing two calculations the manager had decided to combine the two samples and treat them as one sample of size 10 with 3 defectives, what would the posterior probabilities have been?

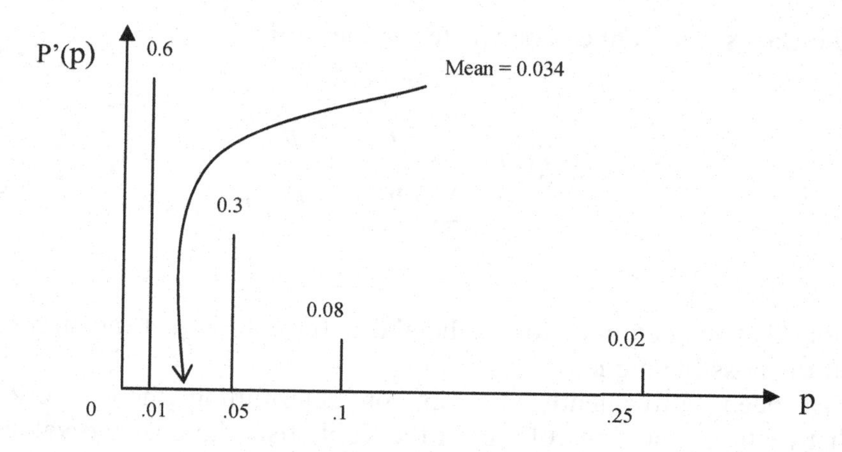


Figure 9.9. Prior distribution of $p$

**Solution**

1. Posterior distribution

Under the assumptions that the production manager has made, the distribution of the number of defectives in 5 trials, given any particular value of $p$, is a binomial distribution. The likelihoods $L(E|p)$ of the evidence E of $k = 1$ defective out of $n = 5$ sampled items are thus:

$$P(k = 1|n = 5, p = 0.01) = \binom{5}{1} \cdot (0.01) \cdot (0.99)^4 = 0.04803$$

$$P(k = 1|n = 5, p = 0.05) = \binom{5}{1} \cdot (0.05) \cdot (0.95)^4 = 0.20362$$

$$P(k = 1|n = 5, p = 0.10) = \binom{5}{1} \cdot (0.1) \cdot (0.90)^4 = 0.32805$$

$$P(k = 1|n = 5, p = 0.25) = \binom{5}{1} \cdot (0.25) \cdot (0.75)^4 = 0.39551$$

The Bayes' theorem can be written in the form:

$$P''(p|E) = \frac{P'(p)P(E|p)}{\sum_p P'(p)P(E|p)}$$

Note that the denominator of the above formula is just the sum of all the possible numerators.
Has been conveniently set up for determining the posterior distribution. The first column of the Table lists the possible values of $p$ (that is the values of $p$ having nonzero prior probabilities). The second and the third columns give the prior probabilities and the likelihoods, and the fourth column is the product of these two columns. Finally, the fifth column is obtained from the fourth column by dividing each element of the column by the sum of the four elements of the same column, i.e. the denominator of the Bayes' theorem in the previous equation. The Figure shows the posterior distribution.

Table 9.5. Step-by-step procedure for applying Bayes' theorem after the first sample

| $p_i$ | $P'(p_i)$ | $L(p_i\|x)$ | $P'(p_i)\cdot L(p_i\|x)$ | $P''(x)$ |
|---|---|---|---|---|
| 0.01 | 0.6 | 0.04803 | 0.028818 | 0.2323 |
| 0.05 | 0.3 | 0.20362 | 0.061088 | 0.4924 |
| 0.1 | 0.08 | 0.32805 | 0.026244 | 0.2115 |
| 0.25 | 0.02 | 0.39551 | 0.00791 | 0.0638 |
| | | | $\sum_i P'(p_i)\cdot L(p_i\|x) = 0.1241$ | 1 |

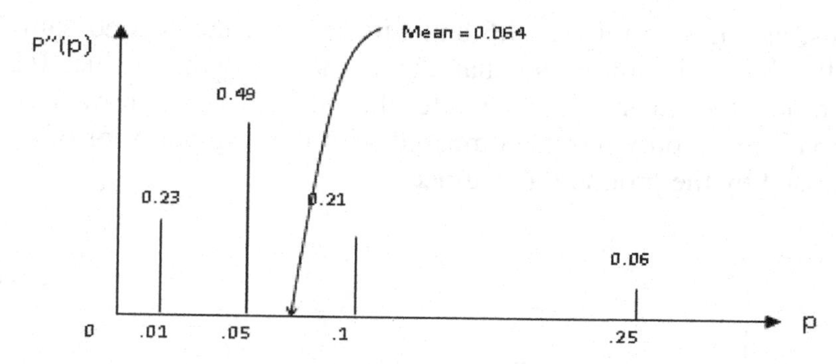


Figure 9.10. Prior distribution of p after the first evidence

Note that, after the first sample, the mean is 0.064, as compared with the prior mean of 0.034. The evidence of one defect out of the 5 sampled items shifts the probabilities towards higher values. The most likely value of the probability moves from 0.01 in the a priori manager's knowledge towards 0.05 after the evidence.

2. Revision of the information inferred in 1. for a new sample.
The step-by-step procedure for revising the probabilities is illustrated in Table 9. The prior information is now the result obtained in (1.) and the likelihoods refer to the event of finding $k$ = 2 defects out of the 5 items. As an example, for the first possible value of $p$:

$$P(k=2|n=5, p=0.01) = \binom{5}{2} \cdot (0.01) \cdot (0.99)^4 = 0.00097$$

Table 9.6. Step-by-step procedure for applying Bayes' theorem after the second sample

| $p_i$ | $P''(p_i)$ | $L(p_i\|x)$ | $P''(p_i) \cdot L(p_i\|x)$ | $P'''(x)$ |
|---|---|---|---|---|
| 0.01 | 0.232 | 0.00097 | 0.000225 | 0.005223 |
| 0.05 | 0.492 | 0.021434 | 0.010546 | 0.244677 |
| 0.1 | 0.212 | 0.0729 | 0.015455 | 0.358575 |
| 0.25 | 0.064 | 0.263672 | 0.016875 | 0.391526 |
| | | | $\sum_i P''(p_i) \cdot L(p_i\|x) = 0.0431$ | 1 |

Thus, as shown in Table 9.6 and Figure 9.11, the second sample shifts the probabilities so that the highest possible value, 0.25, becomes the most likely, while the lowest value now has a probability of only 0.005, compared with the original prior of 0.60 assessed by the production engineer.

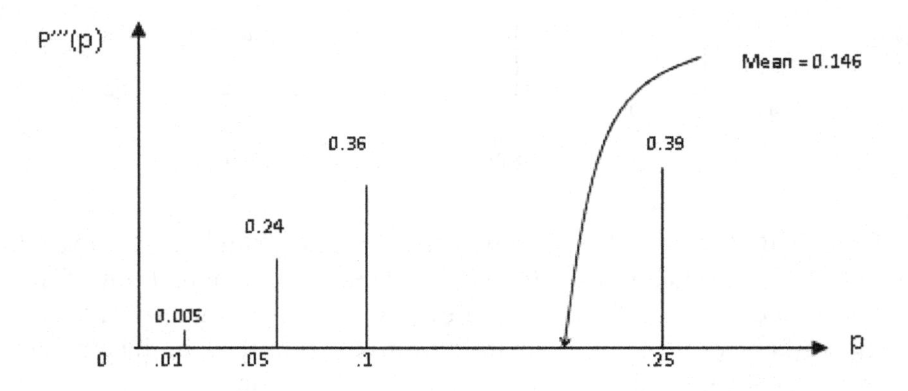


Figure 9.11. Prior distribution of p after the second evidence

3. Results comparison

The information content provided by the evidence does not change if a single sample of 10 items with 3 defectives is considered. Thus we expect, and the theory of Bayes approach confirms it, that the posterior distribution obtained in lights of the evidence of 3 defectives in 10 items is the same as that found in 2.

In this case, the likelihood is the probability of finding $k = 3$ defectives out of $n = 10$ sampled items:

$$P(k = 3|n = 10, p) = \binom{10}{3} \cdot (p)^3 \cdot (1-p)^7 .$$

The step-by-step procedure for updating the prior distribution is reported in Table 9.7. Note that the results obtained for the posterior probability (last column of the Table) are the same of those obtained in 2.. Thus confirming that the same results are achieved either by considering two distinct samples or by condensing the information in a single sample.

Table 9.7. Step-by-step procedure for applying Bayes' theorem when the first and the second samples are taken as one

| $p_i$ | $P'(p_i)$ | $L(p_i \mid x)$ | $P'(p_i) \cdot L(p_i \mid x)$ | $P''(x)$ |
|---|---|---|---|---|
| 0.01 | 0.6 | 0.000112 | 6.71E-05 | 0.00524 |
| 0.05 | 0.3 | 0.010475 | 0.003143 | 0.245377 |
| 0.1 | 0.08 | 0.057396 | 0.004592 | 0.358529 |
| 0.25 | 0.02 | 0.250282 | 0.005006 | 0.390855 |
| | | | $\sum_i P'(p_i) \cdot L(p_i \mid x) = 0.01281\Sigma$ | 1 |

# Chapter 10

# Markov chains

## 10.1 Nuclear steam supply system

A nuclear steam supply system has two turbo-generator units; unit 1 operates and unit 2 is in standby whenever both are good. The units have a constant MTTF of $\lambda_i^{-1}$, $i$=1 and 2, during active operation while during standby unit 2 has a MTTF of $(\lambda_2^*)^{-1}$. The repair of a unit is assumed to begin instantaneously after it fails, but its duration is random so that the instantaneous repair rates will be $\mu_1$ and $\mu_2$, respectively. The repairs can be done on only one unit at a time and any unit under repair will remain so until the task is completed.

1. Draw the system diagram.
2. Write the Markov equations.

### Solution

1. System diagram

Since the system under analysis is characterized by two non identical units, we have five possible states:

0= Unit 1 is in operation and unit 2 is in standby.
1= Unit 1 is under repair and unit 2 is in operation.
2= Unit 2 is under repair and unit 1 is in operation.
3= Both units are down and unit 1 is under repair.
4= Both units are down and unit 2 is under repair.

The state-space diagram becomes:

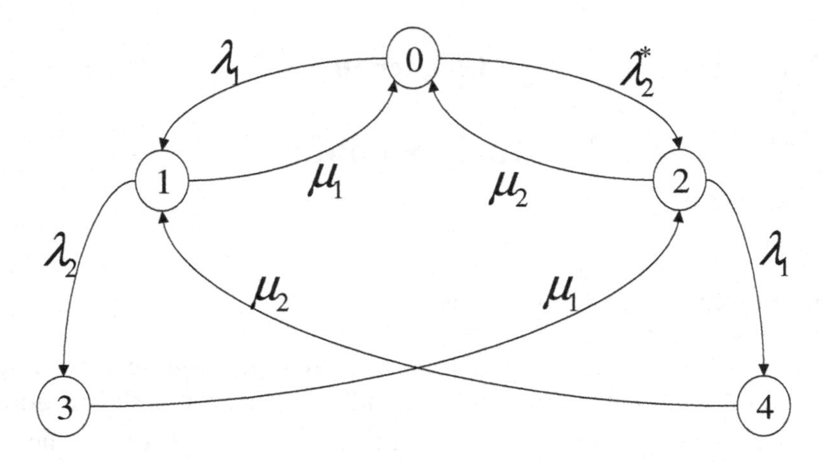


Figure 10.1.

2. Markov equations

Accordingly, the state probabilities equations, written in matrix form, are:

$$\begin{bmatrix}\dot{P}_0(t) & \dot{P}_1(t) & \dot{P}_2(t) & \dot{P}_3(t) & \dot{P}_4(t)\end{bmatrix}$$

$$=\begin{bmatrix}P_0(t) & P_1(t) & P_2(t) & P_3(t) & P_4(t)\end{bmatrix}\cdot\begin{bmatrix}-\lambda_1-\lambda_2^* & \lambda_{11} & \lambda_2^* & 0 & 0\\ \mu_1 & -\mu_1-\lambda_2 & 0 & \lambda_2 & 0\\ \mu_2 & 0 & -\lambda_1-\mu_2 & 0 & \lambda_1\\ 0 & 0 & 0 & -\mu_1 & \mu_1\\ 0 & 0 & 0 & \mu_2 & -\mu_2\end{bmatrix}$$

### 10.2 Alarm system

An alarm system is subject to both unrevealed (u) and revealed (r) faults each of which have time to occurrence which are exponentially distributed with mean values of 200 h and 100 h respectively. If a revealed fault occurs, then the complete system is restored to the time-zero condition by a repair process which has exponentially distributed times to completion with a mean value of 10 h. If an unrevealed fault occurs, then it remains in existence until a revealed fault occurs when it is repaired along with the revealed fault.

1. What is the asymptotic unavailability of the alarm system?
2. What is the mean number of system failures in a total time of 1,000 h?

**Solution**

1. Asymptotic unavailability

We have 3 possible system states:

0= the alarm system is up and no faults have occurred
1= the alarm system is down because a revealed fault has occurred
2= the alarm system is down because an unrevealed fault has occurred

The state-space diagram becomes:

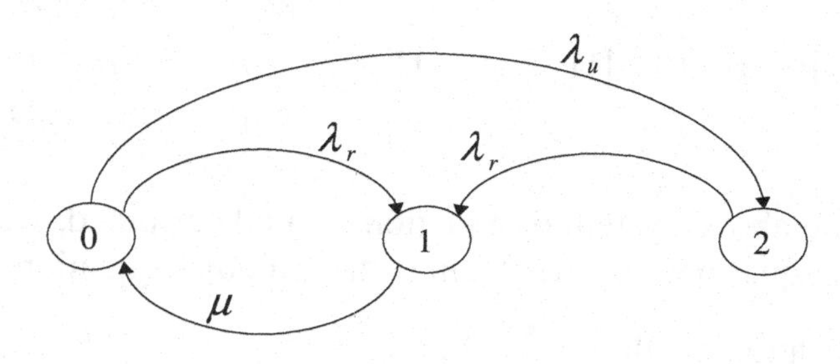


Figure 10.2.

Where

$$\lambda_r = \frac{1}{100}\ \text{h}^{-1} \qquad \lambda_u = \frac{1}{200}\ \text{h}^{-1} \qquad \mu = \frac{1}{10}\ \text{h}^{-1}$$

The corresponding transition matrix $\underline{\underline{A}}$ is:

$$\underline{\underline{A}} = \begin{bmatrix} -\lambda_r - \lambda_u & \lambda_r & \lambda_u \\ \mu & -\mu & 0 \\ 0 & \lambda_r & -\lambda_r \end{bmatrix}$$

and the state equations are:

$$\begin{bmatrix} \dot{P}_0(t) & \dot{P}_1(t) & \dot{P}_2(t) \end{bmatrix}$$
$$= \begin{bmatrix} P_0(t) & P_1(t) & P_2(t) \end{bmatrix} \cdot \begin{bmatrix} -\lambda_r - \lambda_u & \lambda_r & \lambda_u \\ \mu & -\mu & 0 \\ 0 & \lambda_r & -\lambda_r \end{bmatrix}$$

The steady state probability of state *i*, $\Pi_i$, $i = 0,1,2$, is given by $\Pi_i = \lim_{t \to \infty} P_i(t)$. Given that the $\Pi_i$'s do not depend on time, the steady state probabilities can be found by setting to zero the left-hand term of state equations:

$$\begin{bmatrix} 0 & 0 & 0 \end{bmatrix} = \begin{bmatrix} \Pi_0 & \Pi_1 & \Pi_2 \end{bmatrix} \cdot \begin{bmatrix} -\lambda_r - \lambda_u & \lambda_r & \lambda_u \\ \mu & -\mu & 0 \\ 0 & \lambda_r & -\lambda_r \end{bmatrix}$$

Since the above system of equations is undetermined, we solve the set of equations by replacing the middle equation with the normalization condition $\sum_{i=0}^{2} \Pi_i = 1$:

$$\begin{bmatrix} 0 & 1 & 0 \end{bmatrix} = \begin{bmatrix} \Pi_0 & \Pi_1 & \Pi_2 \end{bmatrix} \cdot \begin{bmatrix} -\lambda_r - \lambda_u & 1 & \lambda_u \\ \mu & 1 & 0 \\ 0 & 1 & -\lambda_r \end{bmatrix}$$

this yields:

$$\Pi_0 = \frac{\mu\lambda_r}{\lambda_r^2 + \lambda_r\mu + \lambda_u\mu + \lambda_u\lambda_r} = 0.606$$

$$\Pi_1 = \frac{(\lambda_r + \lambda_u)\lambda_r}{\lambda_r^2 + \lambda_r\mu + \lambda_u\mu + \lambda_u\lambda_r} = 0.303$$

$$\Pi_2 = \frac{\mu\lambda_u}{\lambda_r^2 + \lambda_r\mu + \lambda_u\mu + \lambda_u\lambda_r} = 0.091$$

The asymptotic unavailability $U$ is the probability that the alarm system is down when $t \to \infty$: $U = \Pi_1 + \Pi_2 = 0.394$.

2. Mean number of system failures in a total time of 1000 h
The total frequency of system failures $f$ is the departure rate (frequency) from state 0 to state 1, $\nu_{01}$, plus the departure rate from state 0 to state 2, $\nu_{02}$.

$$f = \nu_{01} + \nu_{02} = \Pi_0 a_{01} + \Pi_0 a_{02} = \Pi_0(\lambda_r + \lambda_u) = 0.00909 \ \text{h}^{-1}$$

where $a_{ij}$ is the *ij*-th element of the transit matrix $\underline{\underline{A}}$. Let then *X* be the random variable denoting the number of system failures occurred over a period of 1000 h; then the expected value of *X*, *E[X]*, is:

$$E[X] = f \cdot 1000 = 9.09$$

### 10.3 Two unit standby system

Consider a two-unit standby system. The failure rate of the operating unit is $\lambda_1$ and that of the standby unit is $\lambda_2$. Both components are operative at time 0.

1. Find the reliability of the system.
2. If, in addition, the system is exposed to a hazard, characterized by a Poisson process with parameter $\lambda_{12}$, that is detrimental to both components, develop an expression for the reliability of the system.
3. Find the *MTTF* in (1) and (2).
4. Assume now that $\lambda_1 = \lambda$, $\lambda_2 = \lambda_{12} = 0$ (cold standby, no common cause failures), and that there is one repairman (repair rate $\mu$). Develop expressions for the steady-state availability, the reliability function, the failure intensity and the mean time to first system failure.

**Solution**

1. Reliability of the system
We have 4 possible system states:

0= both unit are good and the system is up.
1= the operating unit is down, the standby unit is up and so the system is up.
2= the operating unit is up, the standby unit is down and so the system is up.
3= both units are down and the system is down.

The state-space diagram becomes:

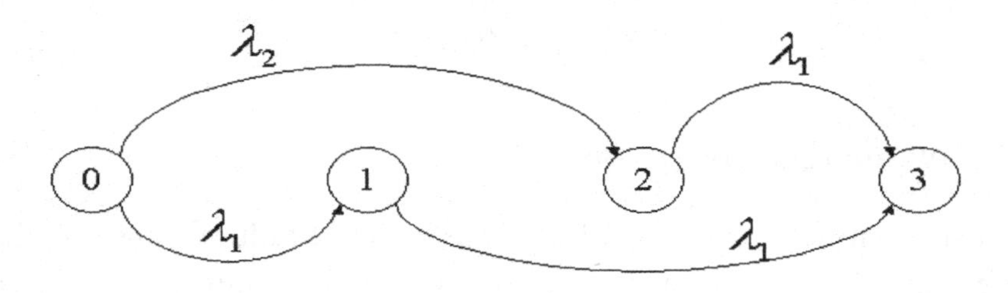


Figure 10.3.

The corresponding transition matrix $\underline{\underline{A}}$ is:

$$\underline{\underline{A}} = \begin{bmatrix} -(\lambda_1+\lambda_2) & \lambda_1 & \lambda_2 & 0 \\ 0 & -\lambda_1 & 0 & \lambda_1 \\ 0 & 0 & -\lambda_1 & \lambda_1 \\ 0 & 0 & 0 & 0 \end{bmatrix}$$

and the state equations are:

$$\lfloor \dot{P}_0(t) \quad \dot{P}_1(t) \quad \dot{P}_2(t) \quad \dot{P}_3(t) \rfloor =$$

$$= [P_0(t) \quad P_1(t) \quad P_2(t) \quad P_3(t)] \cdot \begin{bmatrix} -(\lambda_1+\lambda_2) & \lambda_1 & \lambda_2 & 0 \\ 0 & -\lambda_1 & 0 & \lambda_1 \\ 0 & 0 & -\lambda_1 & \lambda_1 \\ 0 & 0 & 0 & 0 \end{bmatrix}$$

The only state in which the system is down is state number 3, indeed the system reliability is:

$$R(t) = 1 - P_3(t)$$

To find $P_3(t)$ we Laplace-transform the system of state equations above:

$$s\underline{\tilde{P}}(s) - \underline{c} = \underline{\tilde{P}}(s)\underline{\underline{A}}$$

$$\underline{\tilde{P}}(s)(\underline{\underline{A}} - s\underline{\underline{I}}) = -\underline{c}$$

where the vector $\underline{c} = [1 \quad 0 \quad 0 \quad 0]$ indicates the initial condition of the system,

$$\underline{c} = [1 \quad 0 \quad 0 \quad 0]$$

Then, we have:

$$\tilde{\underline{P}}(s) = -\underline{c} \cdot (\underline{\underline{A}} - s\underline{\underline{I}})^{-1} = \underline{c} \cdot (s\underline{\underline{I}} - \underline{\underline{A}})^{-1}$$

Where

$$s\underline{\underline{I}} - \underline{\underline{A}} = \begin{bmatrix} s+\lambda_1+\lambda_2 & -\lambda_1 & -\lambda_2 & 0 \\ 0 & s+\lambda_1 & 0 & -\lambda_1 \\ 0 & 0 & s+\lambda_1 & -\lambda_1 \\ 0 & 0 & 0 & s \end{bmatrix}$$

Then,

$$\tilde{P}_3(s) = \frac{Adj_{03}[s\underline{\underline{I}} - \underline{\underline{A}}]}{Det[s\underline{\underline{I}} - \underline{\underline{A}}]}$$

$$= \frac{\begin{vmatrix} \lambda_1 & \lambda_2 & 0 \\ -(s+\lambda_1) & 0 & \lambda_1 \\ 0 & -(s+\lambda_1) & \lambda_1 \end{vmatrix}}{s(s+\lambda_1)(s+\lambda_1+\lambda_2)} = \frac{\lambda_1(\lambda_1+\lambda_2)}{s(s+\lambda_1)(s+\lambda_1+\lambda_2)}$$

Using the Partial Fractions method we write:

$$\frac{\lambda_1(\lambda_1+\lambda_2)}{s(s+\lambda_1)(s+\lambda_1+\lambda_2)} = \frac{A}{s} + \frac{B}{s+\lambda_1} + \frac{C}{s+\lambda_1+\lambda_2}$$

This yields:

$$\tilde{P}_3(s) = \frac{1}{s} - \frac{\lambda_1+\lambda_2}{\lambda_2(s+\lambda_1)} + \frac{\lambda_1}{\lambda_2(s+\lambda_1+\lambda_2)}$$

We take the inverse Laplace Transform of $\tilde{P}_3(s)$:

$$P_3(t) = 1 - \frac{\lambda_1 + \lambda_2}{\lambda_2} e^{-\lambda_1 t} + \frac{\lambda_1}{\lambda_2} e^{-(\lambda_1+\lambda_2)t}$$

and so

$$R(t) = \frac{\lambda_1 + \lambda_2}{\lambda_2} e^{-\lambda_1 t} - \frac{\lambda_1}{\lambda_2} e^{-(\lambda_1+\lambda_2)t}$$

**10.4 Item with partial repair after first failure**

A new item starts operating on line. When it fails (failure rate $\lambda_1$) a partial repair is performed (repair rate $\mu_P$) which enables the item to continue operation, but with a new failure rate $\lambda_2 > \lambda_1$. When it fails for the second time, a thorough repair (repair rate $\mu_T < \mu_P$) restores the item to the as-good-as-new state and the cycle is repeated.

1. Find the asymptotic unavailability.
2. How can you get the familiar expression for a single item under exponential failure and repair?
3. What is the asymptotic failure intensity?

**Solution**

0 = system up
1= system down, 1st failure
2= system up, partial repair
3= system down, 2nd failure

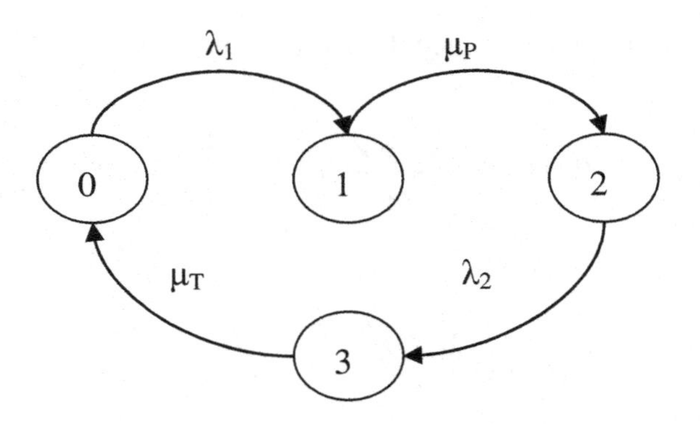


Figure 10.4.

$$\frac{d\underline{P}(t)}{dt} = \underline{P}(t)\underline{\underline{A}} \qquad \underline{\underline{A}} = \begin{bmatrix} -\lambda_1 & \lambda_1 & 0 & 0 \\ 0 & \mu_P & \mu_P & 0 \\ 0 & 0 & -\lambda_2 & \lambda_2 \\ \mu_T & 0 & 0 & -\mu_T \end{bmatrix}$$

Steady States, $\frac{d\underline{\pi}(t)}{dt} = \underline{\pi}(t)\underline{\underline{A}} = 0$

1. Asymptotic unavailability

$$U_\infty = \frac{D_1 + D_3}{\sum D_i} \qquad D_i = \text{determinant}$$

$$D_0 = -\mu_P \lambda_2 \mu_T$$

$$D_1 = -\lambda_1 \lambda_2 \mu_T$$

$$D_2 = -\mu_P \lambda_1 \mu_T$$

$$D_3 = -\lambda_1 \lambda_2 \mu_P$$

$$U_\infty = \frac{\lambda_1 \lambda_2 (\mu_T + \mu_P)}{\mu_T \mu_P (\lambda_1 + \lambda_2) + \lambda_1 \lambda_2 (\mu_T + \mu_P)}$$

2. The familiar expression for a single item under exponential failure and repair

$$U_\infty = \frac{\lambda}{\mu + \lambda}$$

Let $\lambda_1 = \lambda_2 = \lambda$ and $\mu_P = \mu_T = \mu$, then, the unavailability is:

$$U = \frac{2\lambda^2 \mu}{2\mu\lambda(\lambda + \mu)} = \frac{\lambda}{\mu + \lambda}$$

3. Asymptotic failure intensity
Failure intensity = rate of transition out of 'good' states
Intensity = rate of transition ($\lambda$,$\mu$) * probability of state $\pi$
Good states are 0 and 2.
Failure intensity = $\pi_0 \lambda_1 + \pi_2 \lambda_2$

$$W_f(t) = \frac{D_0}{\sum D_i}(\lambda_1) + \frac{D_2}{\sum D_i}(\lambda_2) = \frac{-(\mu_P \lambda_2 \mu_T)}{\sum D}\lambda_1 + \frac{-(\mu_P \lambda_1 \mu_T)}{\sum D}\lambda_2$$

$$W_f(t) = \frac{2\lambda_1 \lambda_2 \mu_T \mu_P}{\mu_T \mu_P(\lambda_1 + \lambda_2) + \lambda_1 \lambda_2 (\mu_T + \mu_P)}$$

### 10.5 Two unit standby system

Consider a two unit standby system, with failure rate $\lambda_a$ and $\lambda_b$ during active operation and $\lambda_b^+$ in the standby mode in which there is a switching failure probability $p$.

1. Draw the transition diagram.
2. Write the Markov equations.
3. Solve for the system reliability.
4. Reduce the reliability to the situation in which the units are identical, $\lambda_a = \lambda_b = \lambda_b^+ = \lambda$.

**Solution**

1. Transition diagram

System states:

0 = A failed, B failed (system failure).
1 = A failed, B functioning.
2 = A functioning, B in stand-by failed.
3 = A functioning, B in stand-by not failed.

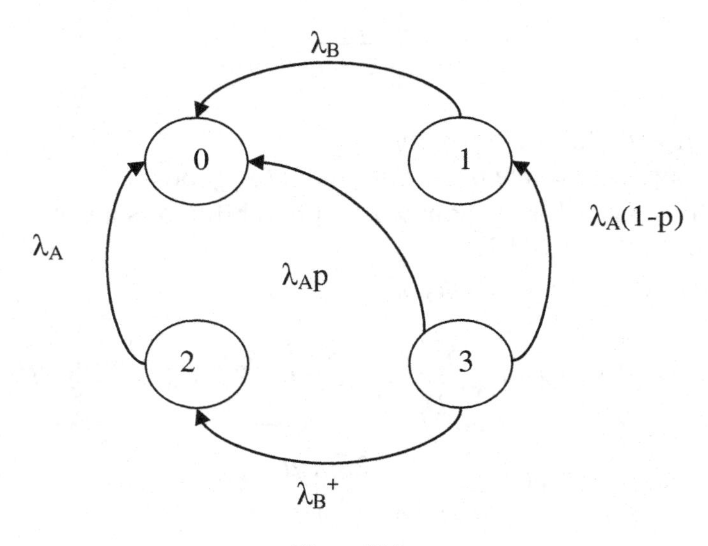


Figure 10.5.

$p$ = probability that the switch has not functioned

2. Markov equations

$$\frac{d\underline{P}(t)}{dt} = \underline{P}(t)\underline{\underline{A}}$$

$$\begin{bmatrix} \dot{P}_0(t) & \dot{P}_1(t) & \dot{P}_2(t) & \dot{P}_3(t) \end{bmatrix}$$

$$= \begin{bmatrix} P_0(t) & P_1(t) & P_2(t) & P_3(t) \end{bmatrix} \begin{bmatrix} 0 & 0 & 0 & 0 \\ \lambda_B & -\lambda_B & 0 & 0 \\ \lambda_A & 0 & \lambda_A & 0 \\ \lambda_A p & \lambda_A(1-p) & \lambda_B{}^+ & -\lambda_A - \lambda_B{}^+ \end{bmatrix}$$

$$P(0) = \begin{bmatrix} 0 & 0 & 0 & 1 \end{bmatrix}$$

3. MTTF of the system

State 0 can be considered as an absorbing state $\Rightarrow$ we obtain the reduced matrix $\underline{\underline{A}}^*$

$$\underline{\underline{A}}^* = \begin{bmatrix} -\lambda_B & 0 & 0 \\ 0 & -\lambda_A & 0 \\ \lambda_A(1-p) & \lambda_B{}^+ & -\lambda_A - \lambda_B{}^+ \end{bmatrix}$$

The system is in state 3 at t=0, so with the Laplace transforms:

$$\underline{\tilde{P}}(0)\underline{\underline{A}}^* = -\underline{c}$$

$$\begin{bmatrix} \tilde{P}_1{}^*(0) & \tilde{P}_2{}^*(0) & \tilde{P}_3{}^*(0) \end{bmatrix} \begin{bmatrix} -\lambda_B & 0 & 0 \\ 0 & \lambda_A & 0 \\ \lambda_A(1-p) & \lambda_B{}^+ & -\lambda_A - \lambda_B{}^+ \end{bmatrix}$$

$$= \begin{bmatrix} 0 & 0 & -1 \end{bmatrix}$$

$$\Rightarrow \begin{cases} -\lambda_B \tilde{P}_1(0) + \lambda_A(1-p)\tilde{P}_3(0) = 0 \\ -\lambda_A \tilde{P}_2(0) + \lambda_B{}^+ \tilde{P}_3(0) = 0 \\ -(\lambda_A + \lambda_B{}^+)\tilde{P}_3(0) = 1 \end{cases} \Rightarrow \begin{cases} \tilde{P}_3(0) = \dfrac{1}{\lambda_A + \lambda_B{}^+} \\ \tilde{P}_2(0) = \dfrac{\lambda_B{}^+}{\lambda_A(\lambda_A + \lambda_B{}^+)} \\ \tilde{P}_1(0) = \dfrac{\lambda_A(1-p)}{\lambda_B(\lambda_A + \lambda_B{}^+)} \end{cases}$$

$$\Rightarrow MTTF = \tilde{R}(0) = \tilde{P}_1 * (0) + \tilde{P}_2 * (0) + \tilde{P}_3 * (0)$$

$$MTTF = \frac{\lambda_A(1-p)}{\lambda_B(\lambda_A + \lambda_B{}^+)} + \frac{\lambda_B{}^+}{\lambda_A(\lambda_A + \lambda_B{}^+)} + \frac{1}{\lambda_A + \lambda_B{}^+}$$

$$MTTF = \frac{\lambda_A{}^2(1-p) + \lambda_B \lambda_B{}^+ + \lambda_A \lambda_B}{\lambda_A \lambda_B(\lambda_A + \lambda_B{}^+)}$$

4. If $\lambda_A = \lambda_B = \lambda_B^+ = \lambda \Rightarrow \qquad MTTF = \dfrac{3-p}{2\lambda}$

**10.6 Water chlorination system**

The water chlorination system of a small town has two separate pipelines, each with a pump that supplies chlorine to the water at prescribed rates. The two pumps are denoted A and B, respectively. During normal operation both pumps are functioning and thus are sharing the load. In this case each pump is operated on approximately 60% of its capacity (cap% = 0.60). When one of the pumps fails, the corresponding pipeline is closed down, and the other pump has to supply chlorine at a higher rate. In this case the single pump is operated at full capacity (cap% = 1.00). We assume that the pumps have the following constant failure rates:
$\lambda_{\text{cap\%}} = \text{cap\%} \cdot 6.3\,\text{y}^{-1}$.

Assume that the probability of common cause failures is negligible. Repair is initiated as soon as one of the pumps fails. The mean time to repair a pump has been estimated to be eight hours, and the pump is put into operation again as soon as the repair is completed. Repairs are carried out independent of each other (i.e., maintenance crew is not a limiting factor). If both pumps are in a failed state, unchlorinated water will be supplied to the customers. Both pumps are assumed to be functioning at time $t=0$.

1. Define the possible system states, and establish a state space (Markov) diagram for the system.
2. Write down the corresponding state equations on matrix format.
3. Determine the steady state probabilities for each of the system states.
4. Determine the mean number of pump repairs during a period of 3 years.
5. Determine the percentage of time exactly when one of the pumps is in a failed state.
6. Determine the mean time to the first system failure (i.e., the mean time until unchlorinated water is supplied to the customers for the first time after time $t=0$).
7. Determine the percentage of time unchlorinated water is supplied to the customers.

**Solution**

1. System state and Markov diagram

Let the number of pumps that are functioning denote the state of the system. The system states are thus {0, 1, 2}. First, we describe the system dynamics by the state-space diagram:

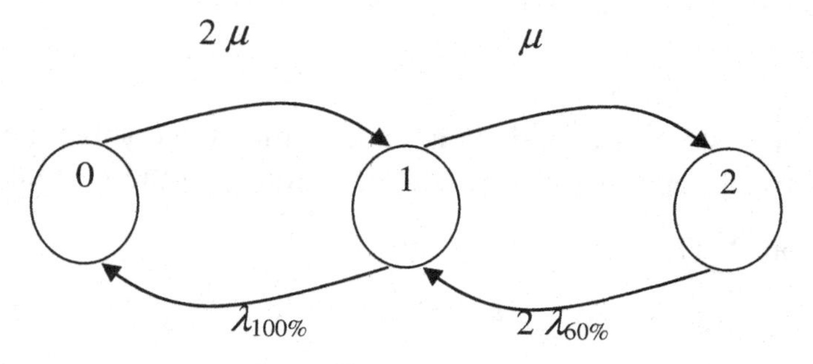


Figure 10.6.

2. State equations

The corresponding transition matrix $\underline{\underline{A}}$ is:

$$\underline{\underline{A}} = \begin{bmatrix} -2\mu & 2\mu & 0 \\ \lambda_{100\,\%} & -\lambda_{100\,\%} - \mu & \mu \\ 0 & 2\lambda_{60\,\%} & -2\lambda_{60\,\%} \end{bmatrix}$$

And the state equations are:

$$\begin{bmatrix} \dot{P}_0(t) & \dot{P}_1(t) & \dot{P}_2(t) \end{bmatrix} = \begin{bmatrix} P_0(t) & P_1(t) & P_2(t) \end{bmatrix} \cdot \begin{bmatrix} -2\mu & 2\mu & 0 \\ \lambda_{100\%} & -\lambda_{100\%} - \mu & \mu \\ 0 & 2\lambda_{60\%} & -2\lambda_{60\%} \end{bmatrix}$$

3. Steady state probabilities

The steady state probability of state *i*, $\Pi_i$, *i*=0, 1, 2, is given by $\Pi_i = \lim_{t \to \infty} P_i(t)$. Given that the $\Pi_i$'s do not depend on time, the steady state probabilities can be found by setting to zero the left-hand term of system state equations:

$$\begin{bmatrix} 0 & 0 & 0 \end{bmatrix} = \begin{bmatrix} \Pi_0 & \Pi_1 & \Pi_2 \end{bmatrix} \cdot \begin{bmatrix} -2\mu & 2\mu & 0 \\ \lambda_{100\%} & -\lambda_{100\%} - \mu & \mu \\ 0 & 2\lambda_{60\%} & -2\lambda_{60\%} \end{bmatrix}$$

Since the system of equations is undetermined, we solve the set of equations by replacing the middle equation with the normalization condition $\sum_{i=0}^{2} \Pi_i = 1$:

$$[0 \quad 1 \quad 0] = [\Pi_0 \quad \Pi_1 \quad \Pi_2] \cdot \begin{bmatrix} -2\mu & 1 & 0 \\ \lambda_{100\%} & 1 & \mu \\ 0 & 1 & -2\lambda_{60\%} \end{bmatrix}$$

This yields:

$$\Pi_2 = \frac{\mu^2}{\mu^2 + 2\mu\lambda_{60\%} + \lambda_{100\%}\lambda_{60\%}} = 0.99312$$

$$\Pi_1 = \frac{2\lambda_{60\%}}{\mu}\Pi_2 = 0.00686$$

$$\Pi_0 = 1 - \Pi_2 - \Pi_1 = 0.00002$$

4. Mean number of pump repairs in 3 years
Repair actions are carried out when the system changes from state 0 to 1 or from state 1 to 2. The total frequency of repairs *f* is hence the departure rate (frequency) from state 0 to state 1, $v_{01}$, plus the departure rate from state 1 to state 2, $v_{12}$.

$$f = v_{01} + v_{12} = \Pi_0 a_{01} + \Pi_1 a_{12} = \Pi_0 2\mu + \Pi_1 \mu = 0.00052$$

where $a_{ij}$ is the ij-th element of the transition rate matrix $\underline{\underline{A}}$. Let then $X$ be the random variable denoting the number of repair actions performed over the period of 3 years; then, the expected value of $X$, $E[X]$, is:

$$E[X] = f \cdot 24 \cdot 365 \cdot 3 = 13.6$$

5. Percentage of time that one pump is in a failed state
At steady state the percentage of time that exactly one pump is in a failed state corresponds to the probability that the system is in state

$$1 = P(\textit{Exactly one of the pumps is failed}) = \Pi_1 = 0.686\%$$

6. Mean time to first system failure
We want to find the time until the first system failure. Given the parallel logic structure, we have to treat state 0 as an absorbing state. This means that all elements in the column corresponding to the absorbing state are set to zero. We also assume that the system is fully functional at time $t = 0$, meaning that the initial state is 2, i.e. $P_2(0) = 1$. The state 0 will "disappear" in the solution of the equations. Therefore, we remove state 0 and its corresponding row and column and obtain a reduced matrix $\underline{\underline{A}}^*$:

$$\underline{\underline{A}}^* = \begin{bmatrix} -\lambda_{100\%} - \mu & \mu \\ 2\lambda_{60\%} & -2\lambda_{60\%} \end{bmatrix}$$

The reduced set of equations can be Laplace-transformed to:

$$s\underline{\tilde{P}}^*(s) - \underline{c} = \underline{\tilde{P}}^*(s)\underline{\underline{A}}^*$$

$$\underline{\tilde{P}}^*(s)(\underline{\underline{A}}^* - s\underline{\underline{I}}) = -\underline{c}$$

The mean time to the first system failure, *MTTF*, can be found from

$$MTTF = \tilde{R}(0) = \tilde{P}_1^*(0) + \tilde{P}_2^*(0)$$

setting $s = 0$,

$$\underline{\tilde{P}}(0)\underline{\underline{A}}^* = -\underline{c}$$

viz.

$$\begin{bmatrix} 0 & -1 \end{bmatrix} = [\tilde{P}_1^*(0) \quad \tilde{P}_2^*(0)] \cdot \begin{bmatrix} -\lambda_{100\%} - \mu & \mu \\ 2\lambda_{60\%} & -2\lambda_{60\%} \end{bmatrix}$$

Solving for $\tilde{P}_1^*(0)$ and $\tilde{P}_2^*(0)$ yields:

$$\tilde{P}_2^*(0) = \frac{\lambda_{100\%} + \mu}{2\lambda_{60\%}\lambda_{100\%}} = 1390$$

$$\tilde{P}_1^*(0) = \frac{2\lambda_{60\%}}{\lambda_{100\%} + \mu}\tilde{P}_2^*(0) = \frac{1}{\lambda_{100\%}} = 202557$$

Finally,

$$MTTF = \tilde{R}(0) = \tilde{P}_1^*(0) + \tilde{P}_2^*(0) = 203948 \text{ hours} \cong 23.3 \text{ years}$$

7. Percentage of time unchlorinated water is supplied to the customers
At steady state the percentage of time that unchlorinated water is supplied to the customers corresponds to the probability that the system is in state 0:

$$P(\text{unchlorinated water is supplied}) = \Pi_0 = 0.002\%$$

### 10.7 Two identical pumps in parallel logic

Two identical pumps are working in parallel logic. During normal operation both pumps are functioning. When one pump fails, the other has to do the whole job alone, with a higher load. The pumps are assumed to have exponentially distributed failure times:

$\lambda_H = 1.5 \cdot 10^{-4}$ h$^{-1}$ when the pumps are bearing 'half load'.
$\lambda_F = 3.5 \cdot 10^{-4}$ h$^{-1}$ when one of the pumps bears the 'full load'.

Both pumps may fail at the same time due to some external stresses. The failure rate with respect to this common cause failure has been estimated to be $\lambda_C = 3.0 \cdot 10^{-5}$ h$^{-1}$. This type of external stress affects the system irrespective of how many of its units are functioning.

Repair is initiated as soon as one of the pumps fails. The mean time to repair a pump, $\mu^{-1}$, is 15 hours. When both pumps are in the failed state, the whole system has to be shut down. In this case, the system will not be put into operation again until both pumps have been repaired. The mean downtime, $\mu_B^{-1}$, when both pumps are failed, has been estimated to be 25 hours.

1. Establish a state-space diagram for the system.
2. Write down the state equation in matrix format.
3. Determine the steady states probabilities.
4. Determine the percentage of time when:

    i. Both pumps are functioning.
    ii. Only one of the pumps is functioning.
    iii. Both pumps are in the failed state.

5. Determine the mean number of pump repairs that are needed during a period of 5 years.
6. How many times we may expect to have a total pump failure (i.e. both pumps in a failed state at the same time) during a period of 5 years?

**Solution**

1. State-space diagram

Let the system be denoted by the number of pumps that are functioning. The system state space is thus {0, 1, 2}. The state-space diagram with corresponding states and transition rates is reported in the following Figure:

System state space:

0 = Both pumps failed.
1 = One pump failed.
2 = Both pumps functioning.

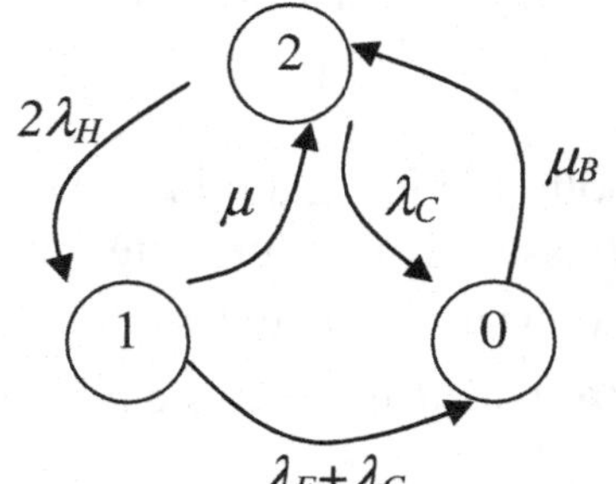

$$\lambda_H = 1.5 \cdot 10^{-4} \text{ h}^{-1}$$
$$\lambda_F = 3.5 \cdot 10^{-4} \text{ h}^{-1}$$
$$\lambda_C = 3.0 \cdot 10^{-5} \text{ h}^{-1}$$
$$\mu = \frac{1}{15} \text{h}^{-1} \qquad \mu_B = \frac{1}{25} \text{h}^{-1}$$

2. Based on the Markov diagram, the equations governing the system dynamics are:

$$\begin{bmatrix} \dot{P}_0(t) & \dot{P}_1(t) & \dot{P}_2(t) \end{bmatrix}$$
$$= \begin{bmatrix} P_0(t) & P_1(t) & P_2(t) \end{bmatrix} \cdot \begin{bmatrix} -\mu_B & 0 & \mu_B \\ \lambda_F + \lambda_C & -\lambda_F - \lambda_C - \mu & \mu \\ \lambda_C & 2\lambda_H & -\lambda_C - 2\lambda_H \end{bmatrix}$$

where $P_i(t)$, $i = 0,1,2$, is the probability of the system being in state $i$ at time t.

3. The steady state probability of state $i$, $\Pi_i$, $i = 0,1,2$, is given by $\Pi_i = \lim_{t \to \infty} P_i(t)$. Given that the $\Pi_i$ do not depend on time, the steady state probabilities can be found by setting to zero the left-hand term of state equations:

$$\begin{bmatrix} \Pi_0 & \Pi_1 & \Pi_2 \end{bmatrix} \cdot \begin{bmatrix} -\mu_B & 0 & \mu_B \\ \lambda_F + \lambda_C & -\lambda_F - \lambda_C - \mu & \mu \\ \lambda_C & 2\lambda_H & -\lambda_C - 2\lambda_H \end{bmatrix} = \begin{bmatrix} 0 & 0 & 0 \end{bmatrix}$$

Since the system of equations is undetermined, we solve the set of equations by replacing the last equation with the normalization condition $\sum_{i=0}^{2} \Pi_i = 1$:

$$[\Pi_0 \quad \Pi_1 \quad \Pi_2]\cdot\begin{bmatrix} -\mu_B & 0 & 1 \\ \lambda_F+\lambda_C & -\lambda_F-\lambda_C-\mu & 1 \\ \lambda_C & 2\lambda_H & 1 \end{bmatrix}=[0 \quad 0 \quad 1]$$

this yields:

$$\Pi_2 = \frac{\mu_B(\lambda_C+\lambda_F+\mu)}{(\lambda_C+\mu_B)(\lambda_C+\lambda_F+\mu)+2\lambda_H(\lambda_C+\lambda_F+\mu)} = 0.99476$$

$$\Pi_1 = \frac{2\lambda_H}{(\lambda_C+\lambda_F+\mu)}\Pi_2 = 0.00445$$

$$\Pi_0 = 1-\Pi_2-\Pi_1 = 0.00079$$

4. The fraction of time $\overline{p_i}$ the system is in state $i$, $i=0,1,2$, over a time T is:

$$\overline{p_i} = \frac{\int_0^T P_i(t)\cdot dt}{T}.$$

When considering the steady states probabilities, $\overline{p_i}(\infty)$ becomes:

$$\overline{p_i}(\infty) = \frac{\int_0^T \Pi_i \cdot dt}{T} = \frac{\Pi_i \cdot T}{T} = \Pi_i$$

Thus, for the three cases:

$\overline{p_2}(\infty) = \Pi_2 = 99.476\%$ is the fraction of time during which both pumps are functioning.

$\overline{p_1(\infty)} = \Pi_1 = 0.445\%$ is the fraction of time during which one pump is functioning.
$\overline{p_0(\infty)} = \Pi_0 = 0.079\%$ is the fraction of time during which both pumps are failed.

5. A repair is performed when the system transfers from state 1 to state 2. Two repairs are performed when the system transfers from state 0 to state 2. The total frequency of repairs, $W_r$, will then be the departure frequency from state 2 to state 1 plus two times the departure frequency from state 0 to state 2, i.e.:

$$W_r = \Pi_1 a_{12} + 2\Pi_0 a_{02} = \Pi_1 \mu + 2\Pi_0 \mu_B = 3.6\ 10^{-4}\ \text{h}^{-1}$$

where $a_{ij}$ is the *ij*-th element of the transit matrix $\underline{\underline{A}}$. Let then *X* be the random variable denoting the number of repair actions performed over the period of 5 years; then the expected value of *X*, *E*[*X*], is:

$$E[X] = W_r \cdot 24 \cdot 365 \cdot 5 = 15.8.$$

6. A total pump failure occurs when the system enters state 0. In a steady state situation the frequency of arrivals (visit frequency) to state 0 is equal to the frequency of departures from state 0 Thus the visit frequency to state 0, $v_0$, is:

$$v_0 = \sum_{j=1}^{2} a_{j0} \Pi_j = a_{00} \Pi_0$$

Let then *Y* be the random variable denoting the number of total pump failures over the period of 5 years, then the expected value of *Y*, E[*Y*], is:

$$E[Y] = v_0 \cdot 24 \cdot 365 \cdot 5 = 1.83$$

# References

Ang, A.H.-S. and Tang, W.H., *Probability Concepts in Engineering Planning and Design, Volume I, Basic Principles*, John Wiley, New York, 1975.

Ang, A.H.-S. and Tang, W.H., *Probability Concepts in Engineering Planning and Design. Volume II, Decision, Risk and Reliability*, John Wiley & Sons, Inc., New York, 1984.

Lewis, E. E., *Introduction to Reliability Engineering*, John Wiley & Sons, New York, 1987, second edition 1995.

Kumamoto, H. and Henley E.J., *Probabilistic Risk Assessment and Management for Engineers and Scientists*, IEEE Press, 1996.

McCormick, N.J., *Reliability and Risk Analysis*, Academic Press, New York, 1981.

Rausand, M. and Høyland, A., *System Reliability Theory: Models, Statistical Methods, and Applications* (2nd ed.), Wiley, Hoboken, 2004.

Ushakov, I.A. and Harrison R.A., *Handbook of Reliability Engineering*, John Wiley & Sons, New York, 1994.

# About the Authors

## ENRICO ZIO

Enrico Zio (BS in nuclear engng., Politecnico di Milano, 1991; MSc in mechanical engng., UCLA, 1995; PhD, in nuclear engng., Politecnico di Milano, 1995; PhD, in nuclear engng., MIT, 1998) is Director of the Chair in Complex Systems and the Energetic Challenge of Ecole Centrale Paris and Supelec, full professor of Reliability, Safety and Risk Analysis at Politecnico di Milano, Rector's delegate for the Alumni Association of the Politecnico di Milano, adjunct professor in Risk Analysis at the University of Stavanger, Norway and at Universidad Santa Maria, Chile, past-Director of the Graduate School of the Politecnico di Milano and invited lecturer and committee member at various Master and PhD Programs in Italy and abroad.

He is the current Chairman of the European Safety and Reliability Association, ESRA, for which he had served in the past as Vice-Chairman (2000-2005), and of the Italian Chapter of the IEEE Reliability Society (2001-).

He has also served as Editor-in-Chief of the International journal Risk, Decision and Policy (2003-2004) and is a member of the Korean Nuclear society and China Prognostics and Health Management society.

He is a member of the editorial board of the international scientific journals Reliability Engineering and System Safety, Journal of Risk and Reliability, Journal of Performability Engineering, Journal of Science and Technology of Nuclear Installations, plus a number of others in the reliability, safety and nuclear energy fields.

He has functioned as Scientific Chairman of three International Conferences and as Associate General Chairman of two others, all in the field of Safety and Reliability.
His research topics are: analysis of the reliability, safety and security of complex systems under stationary and dynamic conditions, particularly by Monte Carlo simulation methods; development of soft computing techniques (neural networks, support vector machines, fuzzy and neuro-fuzzy logic systems, genetic algorithms, differential evolution) for safety, reliability and maintenance applications, system monitoring, fault diagnosis and prognosis, and optimal design.
He is co-author of three international books and more than 170 papers on international journals, and serves as referee of more than 20 international journals.

## PIERO BARALDI

Piero Baraldi (BS in nuclear engng., Politecnico di Milano, 2002; PhD in nuclear engng., Politecnico di Milano, 2006) is assistant professor of Nuclear Engineering at the department of Energy at the Politecnico di Milano. He is the current chairman of the European Safety and Reliability Association, ESRA, Technical Committee on Fault Diagnosis. His main research efforts are currently devoted to the development of methods and techniques for system health monitoring, fault diagnosis, prognosis and maintenance optimization. He is also interested in methodologies for rationally handling the uncertainty and ambiguity in the information. He is co-author of 32 papers on international journals and 38 on proceedings of international conferences.

## FRANCESCO CADINI

Francesco Cadini (BSc in nuclear engineering, Politecnico di Milano, 2000; MSc in aerospace engineering, UCLA, 2003; PhD in nuclear engineering, Politecnico di Milano, 2006) is an assistant professor of Nuclear Engineering at the Politecnico di Milano. His main research efforts are currently devoted to the development and application of *i*) computational methods for the performance assessment of a near surface repository for low level radioactive waste disposal, *ii*) sequential Monte Carlo algorithms (particle filters) for state estimation and prediction with applications to system diagnostic and prognostic and *iii*) soft computing techniques (neural networks, fuzzy logic, genetic algorithms) for model identification, time series prediction and optimal control. He is co-author of more than 40 papers on international journals and proceedings of international conferences.